Lean-AI

We stand at the threshold of a transformative era where the convergence of Lean methodologies and artificial intelligence (AI) is redefining operational excellence. This book is a pioneering exploration of this integration, offering a roadmap for organizations seeking to harness AI's potential while preserving the human-centric values at the heart of Lean. It provides a deep dive into how Lean principles – focused on efficiency, waste reduction, and continuous improvement – can be enhanced through AI-driven intelligence, agility, and innovation. More than just a technical guide, this book is an invitation to rethink the future of work, leadership, and culture in an AI-powered world.

Lean-AI: A Humanistic Integration of Lean and AI empowers readers with the tools and insights necessary to navigate the evolving landscape of Industry 5.0+. Through real-world case studies, practical strategies, and in-depth discussions, this book illustrates how organizations can leverage AI to accelerate Lean transformation, enhance human performance, and drive sustainable growth. Readers will gain actionable knowledge on integrating AI with Lean tools such as 5S, poka-yoke, single minute exchange of dies, total productive maintenance, and value stream mapping, among others. The book also highlights the critical role of leadership and culture in ensuring AI-driven Lean implementations uplift rather than replace human capabilities.

This book is designed for forward-thinking innovators, business leaders, Lean practitioners, and AI enthusiasts eager to explore the synergy between technology and human-centric process improvement. Whether you are an executive driving digital transformation, a Lean expert looking to modernize methodologies, or a professional seeking to understand AI's role in operational excellence, this book provides a strategic guide to mastering Lean-AI integration. It is an essential read for those ready to challenge conventional paradigms and embrace the future of efficiency, collaboration, and intelligent innovation.

Continuous Improvement Series

Series Editors: Elizabeth A. Cudney and Tina Kanti Agustiady

Additive Manufacturing Change Management: Best Practices
David M. Dietrich, Michael Kenworthy, Elizabeth A. Cudney

Healthcare Affordability: Motivate People, Improve Processes, and Increase Performance
Paul Walter Odomirok Sr.

Robust Quality: Powerful Integration of Data Science and Process Engineering
Rajesh Jugulum

Statistical Process Control: A Pragmatic Approach
Stephen Mundwiller

Transforming Organizations: One Process at a Time
Kathryn A. LeRoy

Continuous Improvement, Probability, and Statistics: Using Creative Hands-On Techniques
William Hooper

Affordability: Integrating Value, Customer, and Cost for Continuous Improvement
Paul Walter Odomirok Sr.

Design for Six Sigma: A Practical Approach through Innovation
Elizabeth A. Cudney and Tina Kanti Agustiady

Building a Sustainable Lean Culture: An Implementation Guide
Tina Kanti Agustiady and Elizabeth A. Cudney

Lean Sustainability: A Pathway to a Circular Economy
Elizabeth A. Cudney, Sandra L. Furterer, Chad M. Laux, Gaganpreet S. Hundal

A Career Worth Engineering: Don't Just Graduate – Navigate the Transition from Student to Professional
John S. Rogers and Sean Maciag

Lean-AI: A Humanistic Integration of Lean and AI
Adil Dalal and Elizabeth A. Cudney

For more information about this series, please visit: https://www.crcpress.com/Continuous-Improvement-Series/book-series/CONIMPSER

Lean-AI

A Humanistic Integration of Lean and AI

Adil Dalal and Elizabeth A. Cudney

CRC Press
Taylor & Francis Group
Boca Raton London New York

CRC Press is an imprint of the
Taylor & Francis Group, an **informa** business

Designed cover image: Agus Setyawan © Adil Dalal

First edition published 2026
by CRC Press
2385 NW Executive Center Drive, Suite 320, Boca Raton FL 33431

and by CRC Press
4 Park Square, Milton Park, Abingdon, Oxon, OX14 4RN

CRC Press is an imprint of Taylor & Francis Group, LLC

ISBN: 978-1-032-87503-3 (hbk)
ISBN: 978-1-032-87506-4 (pbk)
ISBN: 978-1-003-53297-2 (ebk)

DOI: 10.1201/9781003532972

Typeset in Times
by Apex CoVantage, LLC

Contents

Foreword 1

The rapid rise of Generative AI (GenAI), which has democratized access to the power of AI's predictive and decisioning capabilities, is revolutionizing every sphere of business. Companies no longer need to invest millions in experimental and unreliable AI solutions that require years to implement and deliver value. With the cloud-based on-tap services of various GenAI platforms, companies can choose and deploy the right model that meets the needs of their use case and business appetite. Barely two years into the advent of GenAI, its industrial applications are multiplying at an incredible pace, offering a plethora of scenarios and options for solving different types of operational and business challenges. And yet GenAI is just an early precursor to the even more powerful and expansive artificial general intelligence (AGI), which is eventually going to subsume every aspect of running a business. The advent of AGI will happen gradually and then suddenly, just as GenAI burst onto everyone's consciousness seemingly out of nowhere. In reality, the groundwork for GenAI was laid over the previous decade, but the progress was incremental until OpenAI's GPT3 was uncloaked in November 2022. Based on a revolutionary transformer model, this large language model (LLM) could interpret regular conversations and churn out human-like responses by crunching terabytes of data in a fraction of a second. It could rapidly answer a plethora of questions or solve complex problems through conversational prompts while also handling multi-modal inputs and outputs producing amazing graphics and videos. Like many of the technologies preceding it, the adoption of GenAI initially started with programmers and then leapfrogged into the realm of the average non-technical consumer of content and media who found a perfect digital companion that could handle multifarious tasks from writing a poem to solving a math homework assignment. The biggest advancement of this disruptive technology is that it provided access to the endless power of hyper-scaled computing fed on petabytes of data to the average person who doesn't have to learn any special skill or programming language other than mastering the art of prompt engineering. GenAI is now advancing into every aspect of one's life and work by unleashing the power of co-pilots and agents that can perform mundane tasks as well as enable synchronous workflows that remove inefficiencies within business functions. This heralds the age of AI where literally every idea can be transformed into action and the art of the possible is within everyone's reach.

Many have been talking about the advent of Industry 5.0, which extends the 4.0 paradigm to add a human edge – a truly autonomous world of seamlessly interconnected devices and machines that can operate end-to-end functions like a worker, but without human intervention. With the convergence between AI, IoT, and automation, this vision is becoming a reality very rapidly. While there is a fundamental dichotomy between the deterministic accuracy of machines and the probabilistic abstractions of AI, by using the right guardrails and frameworks, the two can come together to deliver solutions that are Industry 5.0 level. Quite counter-intuitively, AI can bring the human back into the machines by providing a layer of sentient decisions and actions that can appear to be almost humanoid. The recent emergence of

Agentic AI and Empathic AI is furthering this paradigm shift by taking on roles that think, act, and emote in a way similar to how human beings would react to certain conversations or situations. This opens up the possibility of more human-like machines and bots that don't just follow a calculating prescriptive path but adopt a more human-centered and dynamically adaptive approach to solving problems or finishing tasks. That is one of the wonders of self-learning heuristic AI algorithms, which become smarter as well as "self-aware" through feedback and continuous improvement. The successful use of AI to power a hyper-personalized and digitally seamless future requires that Empathic AI remains an integral element of any AI system. But the biggest challenge to ensuring that AI and humans can work harmoniously is actually in the leadership that decides how to blend them within the value chains. Incorporating AI and machine learning (ML) into machines and processes just for the sake of doing it provides no sustainable benefit. Decisions regarding job displacement, redefining and retooling skill sets, and transparency in complex decision-making are very contextual and impact the business and social outcomes – they are the true indicators of the successful deployment of AI. The age of AI requires a new approach to leadership and design of systems – one that strives for a balance between business and human values. All the tools and methods are at our disposal. It's for us to decide – do we consciously steer toward a more utopian future where AI is seamlessly embedded within our work and lives or do we let technology take control and shape a dystopian reality, which disrupts and distances us from who we are.

Lean and operational excellence is a discipline that has been addressed throughout Industry 2.0 and 3.0. It took a backseat during 4.0 as the general assumption was that we had reached its peak and further productivity gains would be through technology (including automation and analytics). But the advent of AI is opening up a vista of opportunities to not only take Lean to a new level but also eliminate some of its drawbacks and alleviate the human impact. At the surface, Lean with its structured and deterministic approach seems to conflict with AI's inherent traits of probabilistic, ambiguous, and sometimes inaccurate results. But a deeper look reveals how AI could fundamentally fix some of the most significant disadvantages of Lean, such as supply chain risks of just-in-time production or worker burnout and dissatisfaction. Lean and AI can come together to drive productivity and quality significantly higher while humanizing operational excellence by reducing the stress of process rigor and repetitive tasks. Workers who fear the job displacement impact of AI should actually see this as the solution to improving their day-to-day workflow and enhancing the net value of their job functions. AI in fact is the perfect companion for a worker, aiding and performing actions, detecting errors, and pre-empting downstream issues in a value chain.

This book is an eye-opening look into the potential convergence of Lean and AI. It delves into the different aspects and ways of balancing innovation and process discipline to unleash new productivity. It will show business strategists as well as operational executives how to bring Lean and AI together to power next-level innovation and value in any industry. How to mitigate the negative perceptions of AI and instead leverage it as the digital copilot that can improve everyone's work lives.

And very importantly, how to enable sustainability and scalability in an increasingly resource-scarce world. Lean-AI should be required reading for anyone who wants to solve the eternal conundrum of achieving perfection with inspiration, in any field of life.

Anupam Govil
Founder and Chief AI Evangelist, Austin AI Valley
Managing Partner, Avasant

Foreword 2

When asked to give a presentation at a conference, the most common topic I am asked to present is Technology and the Shingo Model. In the 30+ years of administering the Shingo Prize, the Shingo Institute has had a lot of opportunities to see the best organizations in the world.

One thing we have consistently noticed is that one of the big reasons these organizations are the best is because they are the best at adopting new technologies. When comparing these organizations to their competition, we often see they achieve much better productivity with significantly lower investments in technology. It isn't that they are not making investments in technology – they are. But they avoid wasteful investments in technology.

We attribute the difference between this effective use of technology to people and their behaviors. In other words, to their superior culture. The example to all of us is provided by Toyota and the development of *jidōka*. For those unfamiliar with this term, jidōka is the Japanese word for automation. It is written in the kanji as 自動化. Except in Toyota, they created their own word for automation where they substitute the middle character (動), with a different character with the same pronunciation (働). The difference between the standard automation and the Toyota version of automation is called a *ninben* (the addition on the left-hand side of the character) which signifies a person. Hence, Toyota often refers to their version of automation as "automation with a human touch."

The implementation of any technology requires a human touch. Matter of fact, it requires more than just a touch, it requires human involvement. Humans ensure that technology is adopted purposefully and is informed by correct principles.

In the *Shingo Model*, the human aspect of technology is emphasized through the Shingo Guiding Principle of *Respect Every Individual*. Humans decide how technology can make improvements. Humans implement technology in the process. Humans confirm the effectiveness of technology after implementation. The Guiding Principle demands that the role of humans be respected. The Guiding Principles in the Shingo Model are designed to be interactive. For example, respecting every individual *while* improving flow. Dr. Dalal and Dr. Cudney do an excellent job of describing the *Shingo Model*, the role of humans in the improvement process, and how technologies can help humans achieve improvement.

Leaders cannot ignore the potential revolutionary impact of – artificial intelligence (AI). In this book, the authors do an excellent job of tackling what will probably be the most transformative technology of all. They do so by setting the proper context of the purposeful implementation of AI as led by humans. They build on that theme by providing several examples of practical improvements that AI could generate.

Dr. Dalal and Dr. Cudney demonstrate that purposeful adoption of technology can reduce waste, reduce defects, shorten lead times, and improve productivity. One important reason why the best-performing companies in the world are the best is because they are very good at implementing new technologies. The same will be true with AI.

Ken Snyder
Executive Director
Shingo Institute

Preface

We stand at the threshold of a profound transformation. The convergence of Lean principles with artificial intelligence (AI) is not merely an evolution, it is a revolution in how we work, lead, and innovate. *Lean-AI: A Humanistic Integration of Lean and AI* is more than a book; it is a call to action for leaders, practitioners, and visionaries who recognize that the future is not just about automation but about the seamless integration of human ingenuity with technological advancement.

Rooted in efficiency, waste reduction, and continuous improvement, Lean methodologies have long been a cornerstone of operational excellence. Now, combined with AI's dynamic capabilities, they usher in a new era of agility, intelligence, and resilience. While Lean principles provide the strategic compass for culture transformation, AI serves as the engine driving us into uncharted territories of innovation and competence. Together, they redefine what is possible.

This book is a guide for navigating this intricate interplay – one that goes beyond streamlining processes to shaping the very essence of our workplaces. It is about more than just efficiency; it is about how technology influences leadership, collaboration, and the formation of a future-ready, technology-savvy culture.

Both authors of this pioneering work, Adil Dalal and Dr. Beth Cudney, have been at the forefront of Lean and Six Sigma innovation, earning prestigious accolades, including the Shingo Research and Publication Medal and the ASQ Philip Crosby Medal. Their work has transformed industries and shaped the evolution of operational excellence.

AI is not an isolated force but a transformative ally. When guided by ethical and humanistic leadership, AI does not replace human potential – it elevates it. This is the essence of Lean-AI: not just about optimization, but about purpose, culture, and ensuring that AI serves humanity, not the other way around.

AI's influence on cognition, creativity, and development must be handled with care – an imperative that extends beyond individuals to entire organizations. This book ensures that AI enhances the core values of Lean rather than eroding them, fostering workplaces that are not only more innovative and agile but also deeply human-centric.

More than a guide, this book is a manifesto for the next frontier of operational excellence. Through compelling case studies, real-world applications, and a bold vision for Industry 5.0+, we demonstrate how Lean-AI is the key to thriving in the digital era. Expect practical insights, transformative ideas, and moments of revelation as we journey through the landscape of Lean, AI, and the cultural shift that binds them.

If you are ready to challenge the status quo and harness AI to elevate – not replace – human values, join us on this journey. Let's ensure that AI strengthens our principles, enriches our workplaces, and redefines excellence for the future.

Buckle up – Lean-AI is here to revolutionize the way we work, think, and lead.

Authors

Adil Dalal is the President and CEO of Pinnacle Process Solutions, Intl, LLC, holding two master's degrees and a doctorate in business administration and several certifications in leadership, coaching, quality, Lean, Six Sigma, project management, and digital transformations. A visionary leader with 20+ years of global experience, he has driven 350+ enterprise-wide transformations across 25 industries and five continents, spanning Lean Six Sigma, digital transformation, leadership, software development, factory re-layouts, and mergers and acquisitions. A strategic coach to executives and government leaders, Dr. Dalal enhances human potential and return on investment through leadership excellence. Named a Top 15 Entrepreneur Disrupting Their Industry by MSN, he serves on the Forbes Executive Coaching Council and is a Platinum CEO Life Member. A dynamic keynote speaker, he has delivered 100+ talks worldwide, including TEDx Brussels, the Saudi Government's Quality Conference, and the UK House of Lords. His book *The 12 Pillars of Project Excellence* has won the ASQ Crosby Zero-Defect Medal, the Shingo Research Award, and the Axiom Best Business Book Award. Other accolades include the Global Award for Outstanding Contribution to Quality and Leadership, Most Inspiring Thought Leader 2024, and Excellence in Teaching. A pioneer of Sensei Leadership, iLean Technology, iCulture, The V-Factor, and Pinnacle Performance Zone, Dr. Dalal is a catalyst for high-performance teams, revolutionizing leadership, teamwork, and personal development.

Elizabeth A. Cudney, PhD, is President of Cudney Consulting Group, LLC, and a Professor of data analytics at the John E. Simon School of Business at Maryville University. She earned a BS in industrial engineering at North Carolina State University, a master's in mechanical engineering and an MBA at the University of Hartford, and a PhD in engineering management at the University of Missouri–Rolla. Dr. Cudney received the 2025 ASQ Grant Medal for exemplary leadership in designing and delivering curriculum. She received the 2024 Book of the Year Award from the International Lean Six Sigma Institute and the 2023 Walter E. Masing Book Prize from the International Academy for Quality for her book *Lean Sustainability: A Pathway to a Circular Economy*. Dr. Cudney received the 2022 Crosby Medal from ASQ for her book *The Ten Commandments of Lean Six* Sigma. She also received the 2021 Bernard R. Sarchet Award from ASEE EMD for lifetime achievement in engineering management education. She received the 2021 Walter E. Masing Book Prize from the International Academy for Quality for her book *The Ten Commandments of Lean Six Sigma*. In 2018, Dr. Cudney received the ASQ Crosby Medal for her book *Design for Six Sigma: A Practical Approach through Innovation*. She received the 2018 IISE Fellow Award. She also received the 2017 Yoshio Kondo Academic Research Prize from the International Academy for Quality for sustained performance in exceptional published works. In 2014, Dr. Cudney was elected as an ASEM Fellow. In 2013, Dr. Cudney was elected as an ASQ Fellow. In 2010, Dr. Cudney was inducted into the International Academy for Quality. She received the 2008

ASQ A.V. Feigenbaum Medal and the 2006 SME Outstanding Young Manufacturing Engineering Award. She has published 15 books and over 135 peer-reviewed journal papers. Dr. Cudney is a certified Lean Six Sigma Master Black Belt. She holds eight ASQ certifications, which include ASQ Certified Quality Engineer, Manager of Quality/Operational Excellence, and Certified Six Sigma Black Belt, among others.

1 Lean Landscape

HISTORY OF LEAN

Manufacturing has continued to evolve from an individual product crafted by an artisan to highly standardized, yet customizable, mass production. Before the 1900s, products were handcrafted by artisans. Typically, artisans were skilled craftspeople who made each product one at a time from start to completion; therefore, production speed was generally slow, and there were inconsistencies in the final products (Mourtzis & Doukas, 2014). Quality was high because the focus was on that one product. However, there was considerable variation from product to product and in the process because a process step from one product to the following product might be several weeks apart. In addition, artisan training was inconsistent because it largely relied on apprenticeships for training rather than standardized, formal training that was consistent across a field. Therefore, product variation among different artisans was inconsistent. Further, the lack of standardization also meant that the same artisan could not produce the same part repeatability in terms of product dimensions across their products.

Early industrialization in the 1800s and early 1900s introduced mechanized production, which enabled organizations to produce higher volumes of products with lower variation; however, production planning was manual (Marsh, 2012).

In 1913, Henry Ford introduced the moving assembly line, which led to mass production and reduced costs. Machines performed single tasks. One operator was responsible for the process at each machine or station. Therefore, batch production became the dominant production method. This resulted in single-task orientation and more standardized training on specific tasks. Production was high volume, and management focused on high machine utilization. Subsequently, there were large production lots and high inventory. However, the singular task activities resulted in little variation in the process and product (Black & Hunter, 2003). Since the assembly line used batch production, lead times were long because each item in the batch had to wait for all products in that batch to be completed before it continued to the next step in the process (Mikati, 2010).

In the 1950s, Toyota developed the Toyota Production System (TPS), which focused on Just-in-Time (JIT) manufacturing and producing products based on customer demand. TPS also included numerous continuous improvement principles to reduce waste and increase efficiency (Ohno, 1982).

By the 1970s and 1990s, Toyota formalized the Lean manufacturing principles, including methods such as value stream mapping (VSM) and continuous flow to improve production efficiency (Smith & Vidal, 2019). In the 1990s and early 2000s, organizations began integrating Six Sigma and Lean methodologies to improve processes by reducing variation and waste (Gupta et al., 2020). While the Lean methodology began in manufacturing, organizations began applying Lean in service

DOI: 10.1201/9781003532972-1

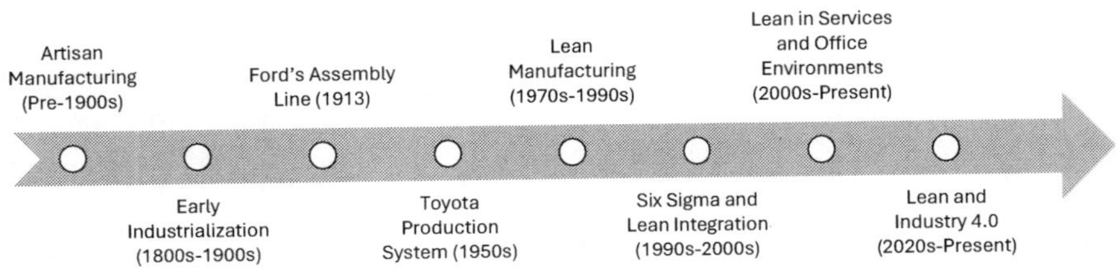

FIGURE 1.1 Evolution of manufacturing systems.

industries, healthcare, finance, and administrative functions to improve operational efficiency.

The Lean principles have continued to evolve to incorporate the principles of Industry 4.0. This has led to the integration of Lean with digital technologies, artificial intelligence (AI), and automation for predictive maintenance, smart manufacturing, and real-time process optimization (Kumar et al., 2021). Organizations use this integrated approach to leverage data analytics and machine learning to further optimize efficiency and decision-making processes.

Figure 1.1 illustrates the timeline of these key developments in manufacturing systems.

THE TRADITIONAL MANUFACTURING ENVIRONMENT

Traditional manufacturing uses verbal production instructions and is still in use today. This type of manufacturing system is based on forecasting and, as a result, requires frequent changes by manufacturing planners. In this environment, organizations produce and schedule production at many locations, with little coordination. Therefore, inventory and materials are decentralized and difficult to control. Production supplies are bought and stored in large quantities based on the forecast. In addition, processes are designed for specialized operators based on skill. Due to the focus on producing to a forecast, the production goals include maximum capacity and machine utilization. Equipment is dedicated to specific functions, resulting in long set-up times for the equipment to justify the large batch-style machine utilization. The lead time to manufacture a product becomes highly variable and complex to predict (Cudney et al., 2013).

THE CONTINUOUS FLOW MANUFACTURING ENVIRONMENT

Continuous flow manufacturing is similar to traditional manufacturing because production includes verbal instructions, is based on forecasted demand, and changes often. Product in continuous flow manufacturing is manufactured at limited locations, and inventory and materials are also still decentralized. The focus of process improvement is now on controlling processes. A major change is that inventory and materials are no longer large, but limited to work-in-process (WIP). In addition, the operator's responsibilities have also changed from specialized tasks to process-related tasks. Equipment is now arranged based on process flow rather than grouped together by function. Finally, product lead times are more predictable, and lead time is reduced (Cudney et al., 2013).

THE STANDARDIZED WORK ENVIRONMENT

In a standardized work environment, production instruction is similar to continuous flow with verbal instructions, forecasted demand, and limited production locations. However, the product is manufactured based on the customer's order with a defined WIP. The work sequence for each process is clearly defined and documented for the operator, which enables flexibility in operator roles rather than requiring specialized skills. The standard work environment enables a seamless movement of operators between machines, regardless of skill level, to perform a specific process function based on the work instructions. In addition, machines are synchronized to approximately the same process speed to reduce bottlenecks and level load the process, and the lead time for a product is predictable (Cudney et al., 2013). Table 1.1 highlights the key characteristics of traditional manufacturing, continuous flow manufacturing, and standardized work environments.

TABLE 1.1
Comparison of Manufacturing Environments

Factor	Traditional Manufacturing	Continuous Flow Manufacturing	Standardized Work Environment
Production instructions	Verbal instructions Production is reactive to demand changes	Verbal instructions Production is forecast-driven	Written instructions Production is based on standardized work instructions
Inventory management	Large, decentralized inventory that is difficult to control	Limited WIP	Defined WIP
Process organization	Individual specialized processes	Processes arranged by flow	Processes are defined and standardized
Lead time	Highly variable and difficult to predict	More predictable with reduced lead times	Predictable and optimized
Operator responsibilities	Specialized, isolated tasks	Related tasks are grouped within processes	Flexible workforce with standardized roles
Equipment utilization	Dedicated to a specific function with long setup times	Arranged by flow to synchronize processes	Machines are synchronized to customer demand
Flexibility	Little flexibility and processes are difficult to adapt	Moderate adaptability	High, seamless substitution of operators

OVERVIEW OF LEAN TOOLS

The Lean methodology includes a comprehensive set of tools to improve efficiency, reduce waste, and foster continuous improvement regardless of industry (Cudney et al., 2013). Lean focuses on driving operational excellence through process improvement initiatives. However, integrating Lean with AI offers even greater potential for optimizing processes. Before discussing the integration of Lean tools with AI, it is important to provide an overview of several key Lean tools.

5S tool is a Lean methodology that focuses on workplace organization. It includes five key components – sort, set in order or straighten, shine or scrub, standardize, and sustain – that aim to create an organized, clean, and safe environment that supports productivity (Hirano, 1995). A sixth "S" is often added to focus on safety. The 6S methodology systematically eliminates clutter, ensures everything is in its proper place, and maintains high standards of cleanliness and safety, which contributes to a more efficient and error-free workspace. The added "S" for safety emphasizes the importance of integrating safety protocols into every process phase to ensure employees work in a secure environment. The integration of 6S and AI is discussed in detail in Chapter 5.

VSM is a foundational Lean methodology used for visualizing a process and identifying process waste. VSM helps organizations analyze the flow of materials and information, which helps to identify bottlenecks, delays, or non-value-added activities that can be eliminated or reduced. By mapping both the current and ideal future state of processes, organizations can use VSM to improve workflows systematically, reduce lead times, and streamline operations, which improves overall efficiency (Rother & Shook, 2003). The integration of VSM and AI is discussed in detail in Chapter 6.

Poka-yoke, or error-proofing, is a Lean tool used in conjunction with root cause analysis and problem-solving methods to prevent defects by addressing potential errors at the source. Poka-yoke minimizes the likelihood of mistakes in a process by using physical barriers, alarms, or sensors, which leads to higher quality and reduced variability. This technique is especially effective in reducing human error because it prevents defects and improves product reliability (Shimbun, 1989). The integration of poka-yoke and AI is discussed in detail in Chapter 7.

Single-minute exchange of dies (SMED) is used in processes that run multiple products or services that include multiple tasks to minimize setup times and changeovers, which are often significant sources of downtime. By distinguishing between internal tasks (performed while machines are stopped) and external tasks (performed while machines are running), SMED streamlines the changeover process and reduces the time from producing the last good piece of one product to the first good piece of the next product. Reducing setup times increases flexibility, which allows organizations to changeover between products quickly and safely and produce smaller batches. The integration of SMED and AI is discussed in detail in Chapter 8.

Total productive maintenance (TPM) is a proactive approach to equipment maintenance that focuses on preventing unplanned downtime (Agustiady & Cudney, 2024). TPM uses planned maintenance, operator involvement, and continuous improvement to ensure that machines are in working condition. By properly maintaining equipment, organizations can extend equipment lifespan, reduce the risk of unexpected breakdowns, and improve overall equipment effectiveness (OEE). TPM leads to improved efficiency, increased productivity, reduced downtime, and reduced operational costs. The integration of TPM and AI is discussed in detail in Chapter 9.

Flow, pull, and kanban are Lean principles that focus on materials management. Flow ensures that products move smoothly through the process by reducing bottlenecks and balancing production steps. Pull systems manufacture products based on actual customer demand rather than forecasts, which reduces overproduction and inventory. Kanban is a visual signaling method that helps manage production by triggering the movement of materials and work based on real-time demand. These tools help reduce excess inventory, eliminate bottlenecks, and keep output aligned with customer needs. The integration of material management methods and AI is discussed in detail in Chapter 10.

Hoshin Kanri, or policy deployment, is a strategic planning tool that aligns an organization's long-term goals with its daily operations. By setting clear objectives and cascading them down through the organization, Hoshin Kanri ensures that every employee works toward the same strategic outcomes (Akao, 1991). Regular reviews and progress checks maintain alignment between operations and overall business strategy, driving continuous improvement and long-term success (Wilson et al., 2024). The integration of Hoshin Kanri and AI is discussed in detail in Chapter 11.

Lean office applies Lean principles to administrative processes to eliminate inefficiencies in office environments. Lean office improves communication and decision-making by streamlining workflows, standardizing procedures, and reducing unnecessary tasks (Martin & Osterling, 2007). This tool is particularly useful in non-manufacturing environments for wasteful practices such as excessive paperwork or redundant approvals. The integration of Lean office and AI is discussed in detail in Chapter 12.

Kaizen and kaikaku are two approaches to process improvement. Kaizen refers to continuous, incremental improvements made by employees at all levels of the organization. Kaizen employs a team-based continuous improvement project or event that focuses on gradual improvements to process efficiency. On the other hand, kaikaku focuses on radical changes and transformative innovations. Kaikaku is often driven by shifts in technology or strategy. Therefore, kaikaku focuses on revolutionary advancements in process efficiency (Bodek, 2004). The integration of continuous improvement approaches and AI is discussed in detail in Chapter 13.

Finally, mixed model production allows organizations to produce multiple variations of a product or multiple different products on the same production line. This method increases flexibility by enabling companies to meet varied customer demands without impacting efficiency. By reducing setup times and producing smaller batches of multiple products, mixed model production helps companies quickly respond to changing market needs while maintaining operational efficiency. Mixed model production is often used in conjunction with other Lean methods such as SMED and standard work. The integration of mixed model production and AI is discussed in detail in Chapter 14.

By understanding these Lean tools, organizations can appropriately apply them to different areas of their operations. When integrated with AI, these tools can be further enhanced. This synergy between Lean methodologies and AI leads to greater process optimization, better decision-making, and sustained competitive advantage.

LIMITATIONS OF LEAN

Lean offers significant benefits in improving efficiency, reducing waste, and fostering continuous improvement; however, it also has several limitations.

Many organizations struggle with implementing Lean because of employee resistance to change and lack of leadership commitment (Antony et al., 2023). Employees may be hesitant to adopt new processes due to the fear of change in established workflows or how it will impact their daily routine. Driving change can be challenging without strong leadership support and clear communication of the benefits of Lean.

In addition, maintaining Lean improvements over time requires continued leadership engagement and ongoing continuous improvement projects. Organizations can struggle to sustain improvements and momentum when leadership focus shifts or employees revert to old habits and potentially lose the gains previously achieved (Turesky & Connell, 2010).

The Lean methodology works best in stable environments with predictable demand. However, in highly volatile markets with rapid changes, organizations often struggle to adapt their processes using Lean (Basu et al., 2021). Therefore, organizations in a dynamic environment must balance Lean principles with flexibility to remain responsive to shifting customer and market needs.

Successful Lean implementation requires ongoing training and commitment at all levels of the organization. Training can be resource-intensive, costly, and time-consuming, especially for organizations with limited resources (Bortolotti et al., 2015). However, ensuring that employees continually develop Lean expertise and have opportunities to apply this knowledge through hands-on projects is essential for sustaining long-term improvements.

While Lean provides tangible benefits such as cost reduction and cycle time improvement, organizations also struggle with measuring intangible benefits such as employee engagement and cultural transformation. It is often a challenge for organizations to quantify improvements in teamwork, communication, and overall workplace morale, which makes it difficult to assess the full impact of Lean initiatives (Alblooshi et al., 2023).

Despite these limitations, organizations can address these challenges through effective leadership, a strong Lean culture, and monitoring Lean initiatives (Antony et al., 2023). Organizations can mitigate these challenges by creating an organizational culture that supports Lean principles and drives sustainable improvements.

LEAN CULTURE

A successful Lean transformation not only depends on implementing various tools but also requires a strong Lean culture throughout the organization that fosters continuous improvement, employee engagement, and leadership commitment (Agustiady & Cudney, 2022).

Leadership commitment is essential for engraining Lean principles into an organization's strategy and culture (Laureani & Antony, 2018). Leaders must actively support and champion Lean initiatives. Their involvement ensures that Lean becomes a fundamental part of the organization's operations rather than a temporary project or viewed as a flavor of the month.

Employee involvement is another critical aspect of a strong Lean culture. A culture of empowerment, where employees are encouraged to contribute ideas and participate in problem-solving, is necessary for sustaining Lean improvements (Neirotti, 2020). Engaged employees take ownership of Lean initiatives and drive continuous improvements.

Continuous learning also plays a significant role in Lean transformation. Organizations must prioritize ongoing training with hands-on projects to develop and build Lean knowledge. Organizations ensure that Lean principles are understood and effectively applied across all levels by providing employees with the necessary knowledge and skills (Reynders et al., 2022).

Standardization and discipline help maintain consistency and reduce variability in processes. Establishing standard work ensures that best practices are followed, which increases efficiency and reliability (Pakdil & Leonard, 2017). Maintaining discipline in Lean practices fosters a structured approach to continuous improvement.

Customer focus is a fundamental principle of Lean. A strong Lean culture emphasizes delivering value to the customer by eliminating waste and improving processes to meet customer expectations (Alves & Alves, 2015). Lean can help organizations enhance service quality, reduce lead times, and improve overall satisfaction by aligning operations and processes with customer needs.

Building and sustaining a Lean culture requires commitment from all levels of the organization and a focus on long-term continuous improvement rather than short-term gains.

CONCLUSION

Lean provides a structured approach to improving operational efficiency, reducing waste, and fostering continuous improvement. Lean has transformed industries through its evolution to modern applications by emphasizing standardization, efficiency, and value creation. However, Lean has challenges, including sustainability issues and difficulty adapting to highly dynamic environments. A strong Lean culture, leadership commitment, and employee engagement are critical factors in overcoming these limitations and ensuring long-term success. Organizations can achieve greater efficiency, enhanced quality, and sustained competitive advantage by understanding and strategically applying Lean principles.

EXERCISES

1. Assess your organization's Lean culture by evaluating leadership commitment, employee involvement, standardization, and continuous learning. Identify areas for improvement.

Lean Culture Factor	Current Status	Suggested Improvement
Example: Leadership commitment	*Moderate*	*Increase Lean training for managers*

2. Discuss strategies for sustaining Lean improvements in your organization. How can leadership ensure that Lean becomes part of the organizational culture?

Sustainability Factor	Strategy
Example: Continuous training	*Monthly Lean workshops*

3. Analyze how Lean aligns with your organization's strategic objectives. Identify key initiatives where Lean principles can drive long-term success.

Strategic Goal	Lean Application
Example: Improve customer satisfaction	*Reduce defects through quality control*

4. Identify key performance indicators (KPIs) that measure Lean effectiveness in your organization. Discuss how metrics can be continuously improved.

KPI	Current Performance	Improvement Strategy
Example: Cycle time	*5 days*	*Reduce through process standardization*

5. Analyzing the role of leadership in successfully implementing and sustaining Lean initiatives. Discuss methods to enhance leadership involvement.

Leadership Factor	Current Involvement	Enhancement Strategy
Example: Predictive maintenance	*Reactive*	*Implement proactive Lean governance*

6. Identify how Lean principles can be used to enhance customer value. Discuss the impact of reducing waste and improving efficiency on customer satisfaction.

Customer Expectation	Lean Improvement
Example: Faster delivery	*Streamline the order fulfillment process*

REFERENCES

Agustiady, T., & Cudney, E. (2022). *Building a sustainable Lean culture: An implementation guide*. New York, NY: CRC Press.

Agustiady, T., & Cudney, E. (2024). *Total productive maintenance: Strategies and implementation guide* (2nd ed.). New York, NY: CRC Press.

Akao, Y. (1991). *Hoshin kanri: Policy deployment for successful TQM*. Productivity Press.

Alblooshi, M., Shamsuzzaman, M., Karim, A., Haridy, S., Shamsuzzoha, A., & Badar, M. A. (2023). Development of a framework for utilising Lean Six Sigma's intangible impacts in creating organisational innovation climate. *International Journal of Lean Six Sigma, 14*(2), 397–428.

Alves, J. R. X., & Alves, J. M. (2015). Production management model integrating the principles of Lean manufacturing and sustainability supported by the cultural transformation of a company. *International Journal of Production Research, 53*(17), 5320–5333.

Antony, J., McDermott, O., Sony, M., Toner, A., Bhat, S., & Cudney, E. (2023). Benefits, challenges, critical success factors, and motivators of quality 4.0: A qualitative global study and directions for future research. *Total Quality Management & Business Excellence, 34*(7–8), 827–846.

Basu, P., Chatterjee, D., Ghosh, I., & Dan, P. K. (2021). Lean manufacturing implementation and performance: The role of economic volatility in an emerging economy. *Journal of Manufacturing Technology Management, 32*(6), 1188–1223.

Black, J. T., & Hunter, S. L. (2003). *Lean manufacturing systems and cell design*. Society of Manufacturing Engineers.

Bodek, N. (2004). *Kaikaku: The power and magic of lean: A study in knowledge transfer*. PCS Inc.

Bortolotti, T., Boscari, S., & Danese, P. (2015). Successful Lean implementation: Organizational culture and soft Lean practices. *International Journal of Production Economics, 160*, 182–201.

Cudney, E., Furterer, S., & Dietrich, D. (2013). *Lean systems: Applications and case studies in manufacturing, service, and healthcare*. New York, NY: CRC Press.

Gupta, S., Modgil, S., & Gunasekaran, A. (2020). Big data in Lean Six Sigma: A review and further research directions. *International Journal of Production Research, 58*(3), 947–969.

Hirano, H. (1995). *5 pillars of the visual workplace*. CRC Press.

Kumar, P., Bhadu, J., Singh, D., & Bhamu, J. (2021). Integration between Lean, Six Sigma and industry 4.0 technologies. *International Journal of Six Sigma and Competitive Advantage, 13*(1–3), 19–37.

Laureani, A., & Antony, J. (2018). Leadership–a critical success factor for the effective implementation of Lean Six Sigma. *Total Quality Management & Business Excellence, 29*(5–6), 502–523.

Marsh, P. (2012). *The new industrial revolution: Consumers, globalization and the end of mass production*. Yale University Press.

Martin, K., & Osterling, M. (2007). *The kaizen event planner: Achieving rapid improvement in office, service, and technical environments*. CRC Press.

Mikati, N. (2010). Dependence of lead time on batch size studied by a system dynamics model. *International Journal of Production Research, 48*(18), 5523–5532.

Mourtzis, D., & Doukas, M. (2014). The evolution of manufacturing systems: From craftsmanship to the era of customisation. In *Handbook of research on design and management of Lean production systems*. New York, NY: IGI Global.

Neirotti, P. (2020). Work intensification and employee involvement in lean production: New light on a classic dilemma. *The International Journal of Human Resource Management, 31*(15), 1958–1983.

Ohno, T. (1982). How the Toyota production system was created. *Japanese Economic Studies, 10*(4), 83–101.

Pakdil, F., & Leonard, K. M. (2017). Implementing and sustaining lean processes: The dilemma of societal culture effects. *International Journal of Production Research, 55*(3), 700–717.

Reynders, P., Kumar, M., & Found, P. (2022). 'Lean on me': An integrative literature review on the middle management role in lean. *Total Quality Management & Business Excellence, 33*(3–4), 318–354.

Rother, M., & Shook, J. (2003). *Learning to see: Value stream mapping to add value and eliminate muda.* Lean Enterprise Institute.

Shimbun, N. K. (1989). *Poka-yoke: Improving product quality by preventing defects.* Crc Press.

Smith, C., & Vidal, M. (2019). The lean labour process: Global diffusion, societal effects, contradictory implementation. In *International handbook of Lean organization* (pp. 150–176). Cambridge: Cambridge University Press.

Turesky, E. F., & Connell, P. (2010). Off the rails: Understanding the derailment of a lean manufacturing initiative. *Organization Management Journal, 7*(2), 110–132.

Wilson, R., Cudney, E., & Marley, R. (2024). Current status of Hoshin Kanri. *The TQM Journal. 36*(2), 460–477.

2 The Emergence of AI and Existing and Future Applications

HISTORY OF ARTIFICIAL INTELLIGENCE

The history of AI is a blend of myth, philosophy, early scientific innovation, and technological breakthroughs. The concept of artificial beings endowed with intelligence has long fascinated humans, with early ideas present in ancient mythologies. For example, Greek mythology mentions mechanical beings, such as Hephaestus's golden automatons, that are capable of thought and speech, which symbolizes humanity's desire to replicate its own cognitive abilities (McCorduck, 2004). As science and philosophy progressed, thinkers such as René Descartes and Gottfried Wilhelm Leibniz laid foundational ideas that foreshadowed AI by questioning the nature of human thought and exploring the mechanization of reasoning. Leibniz, for example, conceptualized a "reasoning calculus" to settle disputes through logic, an early version of the algorithmic reasoning central to AI (Buchanan, 2005).

By the 18th century, early mechanistic inventions, such as Jacques de Vaucanson's mechanical duck and other automota, demonstrated the potential for machines to mimic life. These machines were rudimentary, but they influenced subsequent thinkers and inventors by showing that human actions and thoughts could theoretically be reproduced mechanically. The Industrial Revolution further accelerated interest in mechanization as factories and assembly lines began to automate repetitive tasks, further sparking speculation about the potential for cognitive machines.

AI became a formal scientific field in the mid-20th century based on advances in computational technology. British mathematician Alan Turing posed a revolutionary question: "Can machines think?" Turing proposed the famous "Turing Test" to determine a machine's intelligence based on its ability to exhibit human-like responses, a foundational shift in the quest for automation and the replication of human cognition (McCorduck, 2004; Goh et al., 2006).

The official birth of AI is often traced to the 1956 Dartmouth Conference, where researchers John McCarthy, Marvin Minsky, Nathaniel Rochester, and Claude Shannon explored the possibility of creating intelligent machines (Payette & Abdul-Nour, 2023). This conference coined the term "artificial intelligence" and set the agenda for future research. Early AI research was dominated by symbolic AI or "Good Old-Fashioned AI" (GOFAI), which focused on programming explicit rules and logic to enable machines to solve problems. Researchers developed rudimentary systems capable of basic reasoning, language translation, and even proving theorems,

DOI: 10.1201/9781003532972-2

such as the Logic Theorist by Allen Newell and Herbert A. Simon, which could prove mathematical theorems from *Principia Mathematica*.

Despite initial enthusiasm, AI research hit a bottleneck in the 1970s and 1980s. Researchers identified limitations with AI as they struggled to scale rule-based systems for complex, real-world applications. This period, known as the "AI Winter," was characterized by reduced funding and skepticism due to the failure of AI to meet ambitious promises (Buchanan, 2005).

A paradigm shift occurred in the 1980s and 1990s with the development of machine learning (ML). ML is an approach that enables computers to learn from data instead of relying on fixed rules. Inspired by biological neural networks, researchers created artificial neural networks (ANN) capable of recognizing patterns in data. These neural networks (NN) paved the way for advancements in computer vision, natural language processing (NLP), and other fields where traditional rule-based AI fell short. Algorithms improved the efficiency of NN through approaches such as backpropagation, which enabled machines to recognize complex patterns, such as handwritten characters and facial features (Bashkirova & Krpan, 2024).

The early 21st century marked the beginning of a new era in AI, driven by the availability of large datasets, advanced algorithms, and powerful computational resources. Deep learning is a subset of ML. Deep learning uses multi-layered NN, which enables unprecedented advances in tasks such as image recognition, speech-processing, and game-playing. In 2012, a breakthrough in computer vision occurred when a deep learning model developed by Geoffrey Hinton's team won the ImageNet competition, demonstrating a dramatic improvement in image classification accuracy. This breakthrough began widespread enthusiasm for deep learning, with similar successes in NLP and reinforcement learning. Figure 2.1 illustrates the relationship between AI, ML, deep learning, and ANN.

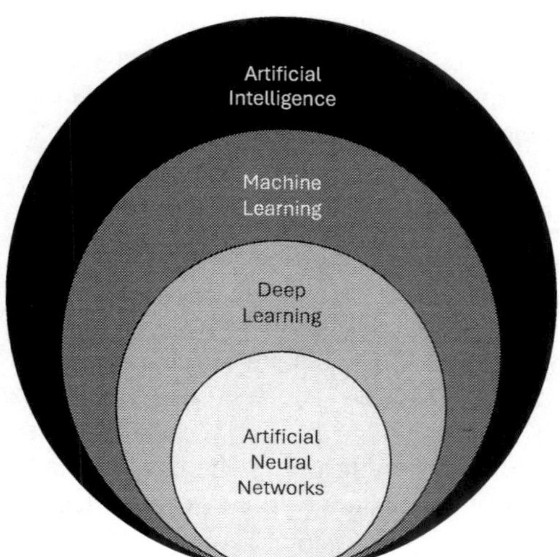

FIGURE 2.1 AI methods.

Reinforcement learning is another approach popularized in recent years. Reinforcement learning teaches AI how to make decisions through trial and error, which is similar to how humans learn from their environment. In 2016, reinforcement learning was demonstrated and received considerable attention when DeepMind's AlphaGo defeated world champion Go player Lee Sedol. This defeat showcased AI's potential for mastering complex strategic tasks. This led to expanding the scope of AI to include applications such as self-driving cars, medical diagnosis, and real-time translation, profoundly impacting industries and society.

AI continues to evolve; however, challenges remain. As models grow increasingly more complex, issues around bias, transparency, and explainability have emerged. AI systems trained on historical data may unintentionally reinforce existing biases, as demonstrated in multiple fields, including criminal justice, hiring, and healthcare (Vicente & Matute, 2023). Furthermore, the "black box" nature of many deep learning models raises concerns about interpretability, particularly in high-stakes applications such as healthcare and finance, where transparency is essential.

Ethical considerations are now central to AI discourse, and researchers are developing frameworks to address fairness, accountability, and transparency issues. Governments worldwide are beginning to establish regulations, such as the European Union's General Data Protection Regulation (GDPR) and the proposed AI Act, to ensure AI is developed and deployed responsibly.

As we look ahead, AI is expected to continue shaping sectors from personalized medicine and environmental monitoring to education and transportation. Research is increasingly focused on creating general-purpose AI, or artificial general intelligence (AGI), that would mimic human intelligence more broadly rather than excel in narrowly defined tasks. This vision is ambitious and fraught with ethical, technical, and philosophical challenges, yet it drives much of the cutting-edge research.

In parallel, the focus has shifted to explainable AI (XAI) in order to make AI systems more transparent and interpretable, allowing users to understand the basis of AI-driven decisions. As AI moves from an experimental technology to a mainstream tool, the focus will shift naturally toward achieving a balance between innovation, safety, and ethical accountability to maximize its benefits for society.

BENEFITS OF AI: SUPPORTING HUMANITY IN NUMEROUS APPLICATIONS

AI is reshaping how industries operate by enhancing productivity, optimizing resource use, and creating new revenue streams. Organizations are using AI to automate repetitive tasks and analyze large datasets, which enables organizations to reduce costs, streamline operations, and make more informed decisions. These advantages are particularly beneficial in finance, healthcare, retail, and manufacturing sectors, where data-driven insights and operational efficiency directly influence profitability.

OPERATIONAL EFFICIENCY

One of AI's most immediate benefits is operational efficiency. By automating labor-intensive tasks, AI minimizes the need for human intervention in repetitive processes, which leads to significant labor cost savings (Mukherjee, 2022). For example, in customer service, AI-powered chatbots and virtual assistants can handle basic inquiries, allowing human agents to focus on more complex issues. This automation reduces labor costs and increases efficiency by enabling faster response times and 24/7 customer support (Casheekar et al., 2024).

In supply chain management, AI-driven predictive analytics allow companies to optimize logistics, inventory levels, and procurement strategies (Nzeako et al., 2024). AI can accurately forecast demand, which reduces excess inventory costs and minimizes stockouts. In manufacturing, predictive maintenance uses AI to analyze equipment data, anticipating when machinery will need repairs. By addressing maintenance issues before they result in breakdowns, companies can prevent costly downtime, extend equipment life, and save on repair costs (Lee et al., 2020).

REVENUE GENERATION AND MARKET EXPANSION

AI is also being used in revenue generation, especially in customer-centric industries such as retail and e-commerce. Companies can analyze consumer behavior, predict purchasing patterns, and personalize marketing through machine learning algorithms (Segun-Falade et al., 2024). For example, e-commerce platforms use AI to recommend products based on a customer's browsing and purchasing history (Zhang & Xiong, 2024). These personalized recommendations increase the likelihood of additional sales, improve customer engagement, and drive revenue growth (Islam et al., 2023).

AI-driven insights enable companies to expand their market reach by identifying new customer segments and customizing products and services for niche markets. For example, streaming services such as Netflix use AI algorithms to analyze viewer preferences, suggesting content that appeals to specific tastes and expanding customer retention and acquisition (Ahmed & Abdulkareem, 2023). Similarly, in financial services, AI is used to identify potential markets for new investment products, assess client risk profiles, and tailor services to meet individual needs, thereby enhancing client satisfaction and loyalty (Ashta & Herrmann, 2021).

PRODUCTIVITY GAINS AND WORKPLACE OPTIMIZATION

AI technology boosts productivity by reducing manual workloads, which enables employees to focus on high-value activities. AI tools can automate routine tasks such as document analysis, contract review, and compliance checks in knowledge-intensive industries, such as legal and financial services (Spring et al., 2022). Legal firms, for example, utilize AI to scan legal documents and identify relevant clauses or precedents, which can drastically reduce the time spent on case preparation (Alarie et al., 2018). This automation speeds up processes and reduces the potential for human error, improving the quality and accuracy of work.

Workforce optimization through AI is also transforming human resources management. AI algorithms can analyze employee performance data, track productivity trends, and provide insights into workforce planning (Tshabalala, 2023). AI-based tools can help organizations identify skill gaps, predict turnover risks, and develop targeted training programs to increase employee engagement and retention (Yanamala, 2024). By aligning workforce capabilities with business goals, AI-driven HR strategies ensure that companies maximize employee potential, which reduces employee turnover and contributes to long-term productivity gains.

ENHANCED DECISION-MAKING AND RISK MANAGEMENT

AI significantly enhances decision-making by giving organizations deep insights into operations, markets, and customer behavior. In finance, for example, AI algorithms analyze massive datasets to detect patterns and anomalies that may indicate risks or opportunities (Kalluri, 2024). These insights allow financial institutions to make data-driven investments, lending, and risk management decisions. AI-powered risk assessment models help banks identify high-risk clients, optimize credit scoring, and assess the likelihood of loan defaults, which leads to better financial outcomes (Edunjobi & Odejide, 2024).

Additionally, AI monitors financial transactions in real time to identify unusual patterns that may signal fraud. By flagging suspicious activities, AI helps prevent fraudulent transactions, which can save financial institutions significant amounts in potential losses (Bello & Olufemi, 2024). This capability also extends to insurance, where AI is used to assess claims, detect fraudulent behavior, and optimize underwriting processes, all of which reduce financial risk and improve profitability.

INNOVATION AND NEW BUSINESS MODELS

AI enables companies to innovate and explore new business models that were previously unattainable or unrealized. For example, in the automotive industry, AI is driving the development of autonomous vehicles, which is a disruptive innovation that can redefine personal transportation, logistics, and ride-sharing services (Torres & George, 2023). By leveraging AI, companies such as Tesla and Waymo are pioneering self-driving technology, which has implications for reducing traffic congestion, lowering emissions, and creating new business opportunities in mobility services (Cascetta & Henke, 2023).

AI also supports innovation in sectors such as healthcare, where it is used to develop personalized medicine. By analyzing genetic data, patient history, and lifestyle factors, AI can help design treatment plans tailored to individual patients, improving health outcomes and reducing healthcare costs. The development of AI-driven drug discovery platforms allows pharmaceutical companies to accelerate research and bring new drugs to market faster, which provides both therapeutic advancements and competitive advantages.

SCALABILITY AND COMPETITIVE ADVANTAGE

AI's scalability allows businesses to grow more efficiently by adapting quickly to market demands. Cloud-based AI solutions make it easier for companies to scale their operations without heavy infrastructure investments, providing small and medium-sized enterprises (SMEs) access to sophisticated tools previously only affordable for large corporations (Hansen & Bøgh, 2021). By adopting AI, companies can respond rapidly to changing customer expectations, refine products based on real-time feedback, and maintain a competitive edge in dynamic markets.

Organizations that integrate AI effectively gain a competitive advantage by becoming more agile, responsive, and data-driven. As AI continues to evolve, the companies that prioritize AI investments are likely to outpace competitors in both operational efficiency and market reach. Moreover, as AI technology becomes increasingly accessible, companies that adopt it early can capitalize on first-mover advantages, securing a dominant position in their respective industries.

CHALLENGES AND STRATEGIC CONSIDERATIONS

While AI offers numerous financial benefits, its implementation requires strategic planning and substantial investment. Companies must account for the costs of developing and maintaining AI infrastructure, data management, and ongoing algorithm updates. Additionally, integrating AI into existing systems often requires organizational restructuring and workforce training, which can be costly and time-consuming.

AI is not a green technology. Its energy consumption is considerable. The computational power required for training and running AI models demands considerable electricity, contributing to high carbon footprints (Zhou et al., 2024). Organizations adopting AI must consider energy efficiency strategies and explore sustainable solutions to mitigate environmental impacts.

Ethical considerations are also important. Companies must ensure AI systems do not inadvertently introduce biases that could lead to discriminatory practices or violate consumer privacy (Li, 2024). For example, using biased historical data in AI models may lead to biased outcomes in hiring, lending, or customer service decisions (Akter et al., 2021). Addressing these challenges requires a commitment to transparent, responsible AI practices aligning with regulatory standards and consumer expectations.

The financial benefits of AI extend across cost reduction, revenue generation, productivity improvements, and enhanced decision-making, with the potential to drive innovation and competitive advantage. As businesses continue to explore and implement AI technologies, understanding these financial advantages and navigating the associated challenges will be critical to maximizing AI's impact on profitability and growth.

COGNITIVE BIASES IN AI-ASSISTED DECISION-MAKING

Cognitive biases are inherent in human thinking and significantly impact how people interact with and perceive AI systems. When humans use AI, biases such as confirmation bias, anchoring bias, and the Dunning-Kruger effect impact their trust, reliance, and judgment. These biases can influence both the design and use of AI, with implications for accuracy, fairness, and effectiveness in high-stakes applications such as healthcare, finance, and criminal justice. Understanding and mitigating these biases is essential for leveraging the potential benefits of AI while minimizing the impact of biases.

CONFIRMATION BIAS

Confirmation bias is the tendency of humans to look for information that supports pre-existing beliefs and plays a prominent role in AI-assisted decision-making. This bias leads users to favor AI outputs that confirm their views, reinforcing initial judgments even if flawed. For example, mental health practitioners may be more inclined to trust AI recommendations that align with their preliminary diagnoses in healthcare. Research suggests that when AI recommendations confirm a practitioner's intuition, they are perceived as more trustworthy and reliable (Bashkirova & Krpan, 2024). However, if an AI's recommendations diverge from a user's expectations, they are often dismissed, preventing practitioners from reconsidering potentially biased initial assessments (Bashkirova & Krpan, 2024).

Confirmation bias can lead to diagnostic errors, especially in high-stakes fields. For instance, an AI used for triage in emergency medicine may suggest prioritizing specific symptoms, but if these recommendations contradict a physician's initial assessment, the physician may ignore critical AI insights, potentially affecting patient outcomes (Bashkirova & Krpan, 2024). In mental healthcare, studies have shown that confirmation bias in diagnosis can result in premature case closure, where practitioners overlook or dismiss contrary evidence, leading to misdiagnosis and suboptimal treatment (Mendel et al., 2011). Addressing confirmation bias requires both training and system design improvements, encouraging users to approach AI outputs with an open mind and carefully consider alternative insights provided by the AI (Bashkirova & Krpan, 2024).

ANCHORING BIAS

Anchoring bias occurs when individuals rely heavily on the first piece of information they encounter when making decisions. This bias can manifest in AI contexts when initial AI recommendations shape subsequent judgments, even if new information contradicts the original suggestion. For example, an AI tool used in financial lending might initially suggest a high-risk profile based on limited data (Tanna & Dunning, 2023). This suggestion could influence a loan officer's perception of an applicant even after additional positive information is available.

This bias can be particularly impactful in environments where decisions evolve as new data becomes available, such as in dynamic supply chain management or credit scoring. When AI provides an initial anchor, users may struggle to adjust their decisions later, even if updated insights reveal a more accurate picture. Designing AI systems to present a broader range of information, rather than a single initial recommendation, could help counteract anchoring bias by encouraging users to consider multiple data points.

AUTOMATION BIAS AND OVER-RELIANCE ON AI

Automation bias refers to the tendency of humans to over-rely on automated systems, assuming that AI is more accurate or capable than it truly is. This tendency can lead users to accept AI recommendations, even when flawed. For example, a study on healthcare decision-making showed that clinicians often trust AI diagnostic tools without cross-verifying AI-generated findings, assuming the AI is objective and free from human error (Vicente & Matute, 2023). This over-reliance can be concerning because it may result in missed errors, especially if the AI system is prone to specific biases or inaccuracies.

Automation bias can have severe consequences in high-stakes areas such as aviation or autonomous driving. For example, in situations where pilots or drivers rely too heavily on autopilot systems, they may fail to notice or react in time to manual override warnings, which could potentially lead to accidents. To mitigate automation bias, designers can create systems that require periodic human intervention or prompt users to confirm AI outputs, ensuring that AI serves as an aid rather than a replacement for human judgment.

THE DUNNING-KRUGER EFFECT

The Dunning-Kruger effect is a cognitive bias in which individuals with limited knowledge overestimate their expertise. It can affect both users and designers of AI systems. Non-experts may exhibit excessive confidence in AI output without understanding the underlying limitations or complexities, leading to misplaced trust. This effect can be particularly problematic in specialized fields such as law or medicine, where non-experts might apply AI insights without fully grasping the context, leading to oversights or even harm.

Conversely, experts in these fields may underestimate the value of AI due to a more nuanced understanding of its limitations. In medical diagnostics, for instance, highly experienced clinicians may approach AI skeptically, relying on their intuition and training over AI outputs. This skepticism can lead to under-utilization of AI, which might otherwise offer valuable support or identify overlooked insights (Salvagno et al., 2023). Addressing the Dunning-Kruger effect requires targeted training for users at all expertise levels, helping them recognize the scope and boundaries of AI capabilities.

ALGORITHMIC BIAS AND TRANSFER OF BIAS

AI models, particularly those trained on historical data, can inherit and even amplify existing societal biases. Algorithmic bias occurs when an AI system reflects the prejudices in its training data. For example, if an AI used in hiring was trained on data from a biased hiring process, it may systematically disadvantage certain demographic groups, perpetuating discrimination. The transfer of bias from AI to human users further complicates this issue, as users may adopt the AI's biased outputs as factual, incorporating these biases into their decision-making processes (Vicente & Matute, 2023).

In judicial settings, for example, predictive policing algorithms have been criticized for reinforcing racial biases by disproportionately targeting minority communities. Studies have shown that users tend to view AI systems as impartial, which can lead them to uncritically accept biased recommendations without questioning their validity. Addressing algorithmic bias involves improving data diversity and model transparency and educating users to critically assess AI recommendations rather than assuming AI outputs are objective (Bashkirova & Krpan, 2024).

STRATEGIES FOR MITIGATING COGNITIVE BIASES IN AI SYSTEMS

Addressing cognitive biases in AI-assisted decision-making requires a multi-faceted approach. First, organizations should prioritize transparency and explainability in AI models to help users understand how AI reaches its conclusions. This approach may involve XAI techniques that clarify the rationale behind an AI's outputs, fostering more informed interactions with the system. XAI helps users see AI recommendations as part of a broader data set, mitigating biases such as confirmation and anchoring by presenting a fuller context for decision-making.

Second, training programs can equip users with the skills to recognize and manage their biases. For instance, in fields such as healthcare and law enforcement, training that includes case studies of cognitive biases in AI use can raise awareness and encourage critical thinking. Additionally, incorporating "checks and balances" into AI systems, such as requiring human validation for high-stakes decisions, can counter automation bias by ensuring that human oversight remains central in decision-making.

Design improvements also play a crucial role in mitigating cognitive biases. AI systems can be designed to provide a range of recommendations rather than a single output, encouraging users to evaluate multiple options rather than anchoring to one suggestion. Periodic prompts reminding users to consider alternative interpretations of the AI's findings or to review key data points independently can further support balanced, bias-aware decision-making.

SYSTEM 1 AND SYSTEM 2 THINKING IN AI: ENHANCING HUMAN COGNITIVE FUNCTIONS

Human cognition is often described in dual-process theories, notably articulated by Nobel laureate Daniel Kahneman in Thinking, Fast and Slow (2011). According to this framework, human thought operates in two distinct modes: System 1 and System 2. System 1 is fast, intuitive, and automatic, while System 2 is slower, deliberate, and analytical. These cognitive systems influence how humans approach problems and make decisions, with System 1 essential for quick judgments in familiar contexts and System 2 for careful analysis in complex, unfamiliar situations. Though fundamentally different from human cognition, AI systems can be designed to emulate aspects of System 1 and System 2, offering insights into how AI can complement human thought processes and enhance decision-making. Figure 2.2 illustrates the differences between System 1 and System 2.

SYSTEM 1 THINKING AND REAL-TIME AI APPLICATIONS

System 1 thinking is characterized by its speed and reliance on instinctual judgments based on pattern recognition. In AI, System 1 functionality can be seen in applications that require immediate, high-frequency responses with minimal human

FIGURE 2.2 System 1 versus System 2.

oversight. For example, AI-driven fraud detection systems operate in real time, continuously scanning financial transactions for anomalies. These algorithms mimic System 1's quick pattern recognition by analyzing vast transactional data to flag suspicious behavior within milliseconds. This ability allows financial institutions to react swiftly to potential fraud without delaying legitimate transactions.

Similarly, AI's role in autonomous vehicles reflects System 1 thinking. Self-driving cars use a variety of sensors and real-time data processing to make instantaneous decisions, such as stopping pedestrians, avoiding obstacles, or responding to sudden changes in traffic flow. These decisions must be made rapidly to ensure passenger safety, and AI's ability to react in real-time mirrors the human capacity for rapid, instinctual responses that System 1 represents. However, just as human System 1 thinking is prone to biases and errors in novel or ambiguous situations, AI systems emulating System 1 may also falter under unpredictable conditions, highlighting the need for thorough testing and fail-safes in high-stakes applications.

SYSTEM 2 THINKING AND ANALYTICAL AI APPLICATIONS

System 2 thinking, by contrast, is more deliberative and methodical, relying on careful analysis rather than instinct. In AI, System 2 processes are represented in applications that require deep data analysis, predictive modeling, and complex decision-making. For example, medical diagnosis AI systems that analyze extensive patient data, genetic information, and medical histories to propose treatment options emulate System 2's thorough and cautious approach. These systems aggregate and process data from diverse sources, applying statistical models to assess risks, predict outcomes, and suggest interventions. In this way, AI can assist medical professionals in making informed, evidence-based decisions by providing a level of analytical depth that is beyond the reach of System 1.

Another example of AI embodying System 2 is in financial portfolio management. AI models analyze market trends, economic indicators, and historical performance data to guide long-term investment strategies. This type of AI supports strategic planning and complex problem-solving, where the stakes are high and rapid decisions are less critical than accuracy and foresight. By emulating System 2, AI assists investors in making informed, data-driven decisions, enabling them to consider various scenarios and assess risks comprehensively.

THE BALANCE BETWEEN SYSTEM 1 AND SYSTEM 2 IN AI

In practice, many AI applications combine System 1 and System 2 thinking elements, adapting their approach depending on the demands of the situation. In healthcare, for instance, AI systems used in radiology often begin with rapid, System 1-like image recognition to detect potential abnormalities in scans. Once flagged, these cases undergo further System 2-like analysis, where the system evaluates the potential diagnosis in detail, weighing various factors to provide a more nuanced assessment. This hybrid approach allows AI to leverage the strengths of both cognitive systems, using quick pattern recognition to identify immediate concerns and detailed analysis to confirm and validate findings.

Similarly, AI uses System 1-like mechanisms in cybersecurity to detect immediate threats, such as unusual login patterns or attempts to breach networks. For more complex security scenarios, however, AI may engage in deeper analysis to identify long-term vulnerabilities, evolving threats, and potential attack patterns, reflecting System 2 thinking. This dual approach enables AI to perform rapid response tasks while also supporting ongoing, strategic security planning.

EINSTEIN'S PROBLEM-SOLVING INSIGHT AND ITS RELEVANCE TO AI

Albert Einstein famously remarked, "If I had an hour to solve a problem, I would spend 55 minutes thinking about the problem and 5 minutes thinking about solutions." This insight aligns closely with System 2 thinking, emphasizing the importance of understanding the problem thoroughly before rushing to action. In AI, this approach is mirrored in the design and training of robust models, where developers must dedicate extensive time to defining the problem, gathering and cleaning data, and choosing appropriate algorithms. In fields such as healthcare and finance, where decisions are consequential, taking time to "think" carefully, as Einstein suggested, can lead to more effective, trustworthy AI systems (Salvagno et al., 2023).

Moreover, AI development benefits from Einstein's philosophy by emphasizing data quality and goal clarity. Machine learning models, for instance, are highly sensitive to the data they are trained on, and inaccuracies or biases in the data can lead to flawed AI outputs. By dedicating time to the data preparation and model evaluation phases, AI practitioners adopt a System 2 mindset, reducing the likelihood of errors and increasing the reliability of the AI's recommendations.

IMPLICATIONS OF SYSTEM 1 AND SYSTEM 2 THINKING FOR AI ETHICS AND ACCOUNTABILITY

The differences between System 1 and System 2 processes in AI raise ethical and accountability considerations. System 1-oriented AI applications, such as autonomous vehicles and automated financial trading, operate with minimal human intervention, leading to questions about accountability when errors occur. Autonomous vehicles, for example, must make split-second decisions, and if a mistake leads to an accident, assigning responsibility is complex. System 2-based AI systems, however, are generally deployed with human oversight, allowing professionals to review AI recommendations before implementing them, as in medical or legal contexts. This aspect allows for greater accountability and safeguards against potential AI errors.

Further, AI operating in System 1 mode must be rigorously tested for bias and fairness since fast, pattern-recognition-based decisions are prone to reproducing biases in training data. This characteristic is particularly critical in high-stakes environments such as hiring, where fast AI screening tools might inadvertently reinforce gender, racial, or socioeconomic biases if not carefully designed and monitored. System 2-based AI systems offer more transparency due to their analytical depth, as they can explain the rationale behind their recommendations. This interpretability is crucial for AI applications in sensitive fields, as it allows users to evaluate AI outputs rather than accepting them unconditionally critically.

TRAINING AND CALIBRATION OF AI FOR EFFECTIVE SYSTEM 1 AND SYSTEM 2 OPERATIONS

Developers and users can enhance AI's effectiveness by calibrating it to balance System 1 and System 2 functions according to context. For example, in environments such as air traffic control, where rapid response is essential but accuracy cannot be compromised, AI systems can be programmed to operate in System 1 mode while maintaining the capacity to shift into a System 2 mode for further verification when uncertainties arise. This adaptive approach can improve the reliability and safety of AI, making it better suited to complex, real-world applications where both speed and accuracy are critical.

Additionally, training users to understand these dual-process modes can improve human-AI collaboration. Professionals in fields such as healthcare and finance can benefit from training that emphasizes when to rely on quick, System 1-like AI outputs, and when to engage System 2-level scrutiny. By understanding the cognitive model behind AI's functionality, users can make more informed decisions, apply critical thinking in scenarios that demand thorough examination, and benefit from rapid AI insights in simpler, time-sensitive situations.

AI STANDARDS: ISO 42001 AND ISO 27001

As AI becomes deeply embedded across industries, standardized guidelines are essential to ensure responsible, secure, and ethical deployment. Two ISO standards – ISO 42001 and ISO 27001 – offer frameworks that address critical aspects of AI management and security. ISO 42001 provides a structured approach for managing AI development and deployment, focusing on ethical considerations and transparency. Meanwhile, ISO 27001 focuses on information security, providing robust measures for safeguarding data integrity, privacy, and compliance, especially crucial in data-intensive AI applications. These standards guide organizations in creating reliable, secure, and ethically grounded AI systems, building public trust in AI technology.

ISO 42001: AI MANAGEMENT SYSTEMS FOR ETHICAL AND TRANSPARENT AI

ISO 42001 is a relatively new standard that provides organizations with a comprehensive framework for managing AI in an ethical, accountable, and transparent way. The standard emphasizes the importance of responsible AI, addressing issues such as fairness, bias mitigation, and transparency. It encourages organizations to establish clear governance policies for AI development and usage, ensuring that AI aligns with core ethical principles and regulatory requirements.

A core component of ISO 42001 is transparency, which requires organizations to maintain detailed documentation of how AI systems operate, including data sources, model training processes, and decision-making logic. By documenting and communicating these elements, ISO 42001 helps prevent AI from becoming a "black box" that users and stakeholders cannot understand or question. For example, in healthcare, where AI systems may influence treatment decisions, transparency allows medical professionals to trust and understand the underlying mechanisms of AI recommendations, fostering greater accountability.

Another fundamental aspect of ISO 42001 is bias mitigation, a critical concern in fields such as hiring, criminal justice, and lending, where AI algorithms can unintentionally reinforce societal biases. The standard requires organizations to assess and mitigate bias at various stages of AI development, from data collection and preprocessing to model training and validation. Bias audits, data diversity checks, and frequent re-evaluations are encouraged to ensure that AI systems make fair and impartial decisions. Implementing ISO 42001 thus helps organizations mitigate the risk of algorithmic bias, promoting fairness and reducing potential discrimination (Bashkirova & Krpan, 2024; Vicente & Matute, 2023).

ACCOUNTABILITY AND ETHICAL OVERSIGHT

ISO 42001 also establishes guidelines for accountability in AI deployment. This standard entails designating individuals or teams responsible for the ethical oversight of AI systems, ensuring that accountability measures are in place at every stage. Organizations are encouraged to set up independent review boards or ethics committees to evaluate the social impact of AI systems, especially in applications where decisions can have significant consequences, such as autonomous vehicles or judicial risk assessments. This approach aligns with the principle of responsible AI, ensuring that human oversight complements AI decision-making.

Moreover, ISO 42001 encourages organizations to engage with stakeholders, including end users, regulatory bodies, and the general public, to align AI systems with societal expectations and norms. Organizations can better anticipate and address potential ethical issues by involving stakeholders in the development and evaluation process. This stakeholder engagement improves trust and aligns AI initiatives with broader societal values and legal requirements.

ISO 27001: INFORMATION SECURITY MANAGEMENT FOR AI

While ISO 42001 addresses AI's ethical and governance aspects, ISO 27001 focuses on the technical and organizational measures required to secure data used in AI systems. As AI systems rely heavily on vast amounts of data, often including sensitive information, ensuring data security is paramount. ISO 27001 provides a robust framework for managing information security risks and protecting against unauthorized access, data breaches, and loss of data integrity.

ISO 27001 requires organizations to conduct thorough risk assessments to identify vulnerabilities in their data infrastructure, establish access controls, and implement data encryption practices. This requirement means securing data pipelines, model training environments, and data storage for AI systems to protect sensitive information. For example, healthcare AI applications that process patient data must comply with privacy regulations such as the Health Insurance Portability and Accountability Act (HIPAA) in the United States, and ISO 27001 offers the necessary tools and protocols to align with such regulations, ensuring data privacy and security (Salvagno et al., 2023).

Another critical aspect of ISO 27001 is incident management, which establishes protocols for responding to security breaches and data leaks. In AI, where models often rely on real-time data, a breach can compromise the integrity of AI outputs. By establishing a clear incident management framework, organizations can quickly detect, respond to, and mitigate security incidents, minimizing damage and maintaining the reliability of AI systems. This aspect of ISO 27001 is essential in sectors such as finance and healthcare, where data security breaches can have severe legal, financial, and reputational consequences.

RISK MANAGEMENT AND COMPLIANCE WITH REGULATORY REQUIREMENTS

ISO 27001 also emphasizes risk management in data handling and AI deployment, encouraging organizations to regularly assess and mitigate potential security risks. This standard includes identifying possible threats, assessing the likelihood and impact of each threat, and implementing security controls to address identified vulnerabilities. For AI systems, this could mean monitoring data inputs to detect anomalous activity, ensuring that training datasets are not corrupted, and regularly auditing AI models to ensure compliance with security standards.

In addition, ISO 27001 enables organizations to comply with data protection regulations, such as the General Data Protection Regulation (GDPR) in the European Union, which mandates strict data handling, storage, and privacy protocols. By aligning with ISO 27001, organizations can demonstrate compliance with GDPR and other regional privacy laws, reducing the risk of penalties and enhancing trust with users and stakeholders.

SYNERGIES BETWEEN ISO 42001 AND ISO 27001 FOR COMPREHENSIVE AI GOVERNANCE

ISO 42001 and ISO 27001, while addressing different aspects of AI management, complement each other to provide a holistic framework for responsible AI deployment. ISO 42001 focuses on ethical governance and bias mitigation, while ISO 27001 ensures data security and privacy compliance. Together, these standards help organizations build trustworthy AI systems that are both ethically sound and technically secure.

For example, an AI application in the hiring process may leverage ISO 42001 to implement unbiased decision-making protocols while using ISO 27001 to secure candidate data. In healthcare, ISO 42001 can guide organizations in ethically deploying diagnostic AI tools, while ISO 27001 ensures that patient data remain secure. By implementing both standards, organizations create a layered approach to AI governance that addresses ethical considerations and technical safeguards, ensuring that AI applications responsibly and securely serve societal needs.

CHALLENGES IN IMPLEMENTING ISO 42001 AND ISO 27001

Despite their benefits, implementing ISO 42001 and ISO 27001 can be challenging. Both standards require significant time and resources, including establishing new governance structures, extensive documentation, and ongoing monitoring. Smaller organizations, in particular, may find it challenging to meet these requirements without substantial technical and financial support. Moreover, as AI systems continue to evolve, organizations may need to continuously update their compliance efforts to address emerging risks and changes in regulatory requirements.

ISO 42001 also presents challenges in defining and measuring ethical criteria, as concepts such as fairness and accountability can be subjective and vary across cultural or regional contexts. Similarly, maintaining ISO 27001 compliance demands constant vigilance against new cybersecurity threats, requiring organizations to adopt an agile approach to information security that can evolve with the changing digital landscape.

FUTURE DIRECTIONS FOR ISO STANDARDS IN AI

As AI advances, ISO standards will likely evolve to address emerging ethical, social, and technical challenges. Future iterations of ISO 42001 may include more specific guidelines on transparency and explainability, given the growing importance of XAI in building user trust. Additionally, ISO 27001 may expand to address the unique cybersecurity risks associated with AI, such as adversarial attacks where malicious actors manipulate AI systems by altering inputs. These adaptations will ensure that ISO standards remain relevant and robust in guiding AI governance and security.

The evolution of these standards will likely be influenced by advancements in AI policy globally. Countries are increasingly developing AI-specific regulations, such as the European Union's proposed AI Act, which could set precedents for ISO updates. Organizations can ensure that their AI systems meet the highest ethical and security standards by aligning with national regulations and global standards, fostering a sustainable and trustworthy AI ecosystem.

FUTURE APPLICATIONS OF AI

AI technologies are poised to revolutionize numerous sectors, bringing advancements in healthcare, sustainability, education, transportation, and finance while also raising important ethical and regulatory questions.

AI in healthcare is advancing precision medicine and predictive diagnostics, enabling tailored treatments and early disease detection. By analyzing large datasets, AI can identify genetic markers for targeted therapies and detect diseases such as cancer or Alzheimer's early, improving patient outcomes. In mental health, AI-powered chatbots offer accessible support, helping to address growing demand in regions with limited mental health resources. However, concerns about data privacy and accuracy in diagnosis remain significant.

AI will play a critical role in climate change mitigation and sustainable practices. In energy management, AI optimizes renewable resources, enhancing grid efficiency and reducing fossil fuel reliance. Precision agriculture, supported by AI, allows for more sustainable water and fertilizer usage, while AI-powered biodiversity monitoring aids conservation efforts. Ethical considerations include data ownership and equitable access to these technologies.

AI-powered platforms in education offer personalized learning experiences, adapting content to each student's pace and needs. Virtual tutors provide 24/7 support, giving students instant assistance in subjects such as math and language. While AI holds promise for improving educational outcomes, issues of data privacy and equitable access persist, particularly in under-resourced regions.

Autonomous vehicles (AVs) and smart city infrastructure will reshape urban mobility. AVs can enhance road safety and accessibility, while AI-optimized traffic management in smart cities reduces congestion and emissions. These innovations raise questions about safety, data privacy, and accountability, particularly in the event of accidents or system failures.

AI is transforming finance through enhanced risk management, fraud detection, and personalized investment strategies. AI analyzes vast datasets to identify financial risks, improving credit scoring and fraud prevention. AI-driven trading also adapts to real-time market trends. Ensuring transparency and preventing algorithmic bias are critical as AI becomes more embedded in financial decision-making.

AI in media will deliver highly personalized experiences. Streaming platforms use AI for tailored recommendations, while AI in VR and AR enables responsive, interactive environments. However, ethical concerns around privacy and the psychological impact of immersive AI experiences are growing.

AI's future applications can have wide-ranging societal benefits. However, they must be developed by carefully considering ethical and regulatory implications. Responsible AI deployment is essential to ensure it has a positive impact on diverse sectors, addressing potential risks while enhancing societal well-being.

CONCLUSION

AI has quickly progressed from theoretical exploration to powerful applications that impact industries and daily life. AI's capabilities to automate complex processes, enhance decision-making, and drive operational efficiency have proven invaluable, generating substantial financial benefits and enabling new business models across fields. However, understanding its practical and ethical implications is essential as AI becomes more widely used.

Historically, AI has been a continuous evolution, marked by technological breakthroughs and shifting paradigms. Early systems that relied on rigid rules have given way to more adaptable, data-driven models. This has enabled AI to address complex tasks such as medical diagnostics, financial forecasting, and personalized customer experiences. However, as AI becomes more autonomous and influential, the concerns surrounding cognitive biases, data privacy, and transparency become increasingly relevant. Cognitive biases, including confirmation bias and the Dunning-Kruger effect, can affect AI interactions, highlighting the need for design strategies that mitigate these biases and promote balanced human-AI collaboration.

Standards such as ISO 42001 and ISO 27001 provide frameworks for ethical governance and data security for AI deployment. ISO 42001 encourages fairness and accountability, while ISO 27001 focuses on protecting data integrity and privacy, which is especially important in sensitive applications such as healthcare and finance. These standards support the development of trustworthy, secure AI systems that meet societal expectations and regulatory requirements.

AI's future potential spans all fields, from healthcare to education and environmental sustainability. Each field offers opportunities to address critical challenges and improve quality of life. However, with these advances come responsibilities to ensure AI is implemented ethically, transparently, and responsibly. Balancing innovation and accountability will be key to using AI for positive change and fostering a future where technology serves humanity's broader goals of sustainability, inclusivity, and progress.

EXERCISES

1. Identify the tactical, operational, and strategic benefits of integrating AI into your current processes.

Process Area	Tactical Benefits	Operational Benefits	Strategic Benefits
Example: Maintenance Process	*Faster issue detection and reduced downtime*	*Optimized preventive maintenance scheduling*	*Extended equipment lifespan, lower long-term costs*

2. Analyze the current limitations in your processes and how AI can address them.

Process Area	Current Limitations	AI Approach to Overcome Limitations
Example: Inventory Management	*Inaccurate stock levels, human errors*	*AI-powered automated inventory tracking and optimization*

3. Develop a strategy for integrating AI into the tactical, operational, and strategic levels of your processes.

Process Area	Tactical AI Integration Strategy	Operational AI Integration Strategy	Strategic AI Integration Strategy
Example: Quality Control	*Implement AI-driven defect detection*	*Use AI for real-time process adjustments*	*Leverage AI for long-term quality trend analysis*

4. Evaluate the potential ethical challenges and data privacy concerns in implementing AI into your processes.

Ethical or Privacy Concerns	Impact on Workforce or Organization	Proposed Solutions
Example: Employee Surveillance	*Reduced morale due to constant monitoring*	*Implement transparency and clear data usage policies*

5. Reflect on the short- and long-term impacts of AI integration on your organization.

Area of Impact	Short-term Impact	Long-term Impact
Example: Productivity	*Immediate increase in efficiency*	*Sustainable growth in overall process performance*

6. Consider the technology requirements for AI integration and develop strategies for successful implementation.

Technology Required for AI	Integration Strategy
Example: Predictive maintenance	• *Integrate with existing equipment* • *Regular updates to AI algorithms*

7. Develop a high-level roadmap that outlines the steps necessary for successfully integrating AI with Lean tools. Include phases such as readiness assessment, technology selection, pilot testing, scaling, and sustaining improvements.

Phase	Key Actions
Example: Readiness assessment	*Identify processes that would benefit most from AI*
Technology Selection	
Pilot Testing	
Scaling	
Sustaining Improvements	

8. Identify specific areas within your Lean practices where AI could enhance operational efficiency, improve decision-making, or reduce waste. Consider AI applications in tools such as 5S, kanban, or VSM.

Lean Practice	Current Challenge	AI Opportunity
Example: 5S	*Inconsistent workplace organization*	*Use AI-powered sensors to monitor and provide real-time alerts*

9. Reflect on the ethical challenges your organization may face when integrating AI with Lean tools. Consider how you will address concerns about employee privacy, decision-making autonomy, and data security. Propose strategies to handle these challenges responsibly.

Ethical Concern	Potential Impact	Proposed Strategy
Example: Employee privacy	*Risk of reduced employee trust due to AI monitoring*	• *Foster transparency* • *Involve employees in AI implementation decisions*

REFERENCES

Ahmed, A., & Abdulkareem, A. M. (2023). Big data analytics in the entertainment industry: Audience behavior analysis, content recommendation, and revenue maximization. *Reviews of Contemporary Business Analytics*, *6*(1), 88–102.

Akter, S., McCarthy, G., Sajib, S., Michael, K., Dwivedi, Y. K., D'Ambra, J., & Shen, K. N. (2021). Algorithmic bias in data-driven innovation in the age of AI. *International Journal of Information Management*, *60*, 102387.

Alarie, B., Niblett, A., & Yoon, A. H. (2018). How artificial intelligence will affect the practice of law. *University of Toronto Law Journal*, *68*(Supplement 1), 106–124.

Ashta, A., & Herrmann, H. (2021). Artificial intelligence and fintech: An overview of opportunities and risks for banking, investments, and microfinance. *Strategic Change*, *30*(3), 211–222.

Bashkirova, A., & Krpan, D. (2024). Confirmation bias in AI-assisted decision-making: AI triage recommendations congruent with expert judgments increase psychologist trust and recommendation acceptance. *Computers in Human Behavior: Artificial Humans*, *2*(1), 100066.

Bello, O. A., & Olufemi, K. (2024). Artificial intelligence in fraud prevention: Exploring techniques and applications challenges and opportunities. *Computer Science & IT Research Journal*, *5*(6), 1505–1520.

Buchanan, B. G. (2005). A (very) brief history of artificial intelligence. *AI Magazine*, *26*(4), 53–60.

Cascetta, E., & Henke, I. (2023). The seventh transport revolution and the new challenges for sustainable mobility. *Journal of Urban Mobility*, *4*, 100059.

Casheekar, A., Lahiri, A., Rath, K., Prabhakar, K. S., & Srinivasan, K. (2024). A contemporary review on chatbots, AI-powered virtual conversational agents, ChatGPT: Applications, open challenges and future research directions. *Computer Science Review*, *52*, 100632.

Edunjobi, T. E., & Odejide, O. A. (2024). Theoretical frameworks in AI for credit risk assessment: Towards banking efficiency and accuracy. *International Journal of Scientific Research Updates*, *7*(1), 92–102.

Goh, O. S., Wong, K. W., Fung, C. C., & Depickere, A. (2006). Towards a more natural and intelligent interface with embodied conversation agent. In *2006 international conference on Game research and development*. Murdoch University.

Hansen, E. B., & Bøgh, S. (2021). Artificial intelligence and internet of things in small and medium-sized enterprises: A survey. *Journal of Manufacturing Systems*, *58*, 362–372.

Islam, M. R., Shawon, R. E. R., & Sumsuzoha, M. (2023). Personalized marketing strategies in the US retail industry: Leveraging machine learning for better customer engagement. *International Journal of Machine Learning Research in Cybersecurity and Artificial Intelligence*, *14*(1), 750–774.

Kahneman, D. (2011). *Thinking, fast and slow*. Farrar, Straus and Giroux.

Kalluri, K. (2024). AI-driven risk assessment model for financial fraud detection: A data science perspective. *International Journal of Scientific Research and Management*, *12*(12), 1764–1774.

Lee, J., Ni, J., Singh, J., Jiang, B., Azamfar, M., & Feng, J. (2020). Intelligent maintenance systems and predictive manufacturing. *Journal of Manufacturing Science and Engineering*, *142*(11), 110805.

Li, Z. (2024). Ethical frontiers in artificial intelligence: Navigating the complexities of bias, privacy, and accountability. *International Journal of Engineering and Management Research*, *14*(3), 109–116.

McCorduck, P. (2004). *Machines who think: A personal inquiry into the history and prospects of artificial intelligence* (2nd ed.). A K Peters/CRC Press.

Mendel, R., Traut-Mattausch, E., Jonas, E., Leucht, S., Kane, J. M., Maino, K., Kissling, W., & Hamann, J. (2011). Confirmation bias: Why psychiatrists stick to wrong preliminary diagnoses. *Psychological Medicine*, *41*(12), 2651–2659.

Mukherjee, A. N. (2022). Application of artificial intelligence: Benefits and limitations for human potential and labor-intensive economy–an empirical investigation into pandemic ridden Indian industry. *Management Matters*, *19*(2), 149–166.

Nzeako, G., Akinsanya, M. O., Popoola, O. A., Chukwurah, E. G., & Okeke, C. D. (2024). The role of AI-driven predictive analytics in optimizing IT industry supply chains. *International Journal of Management & Entrepreneurship Research*, *6*(5), 1489–1497.

Payette, M., & Abdul-Nour, G. (2023). Machine learning applications for reliability engineering: A review. *Sustainability*, *15*(7), 6270.

Salvagno, M., Caravagna, G., & Bergamaschi, S. (2023). AI's role in financial decision-making and transparency. *Critical Care*, *27*(1), 180. https://doi.org/10.1186/s13054-023-04473-y

Segun-Falade, O. D., Osundare, O. S., Kedi, W. E., Okeleke, P. A., Ijomah, T. I., & Abdul-Azeez, O. Y. (2024). Utilizing machine learning algorithms to enhance predictive analytics in customer behavior studies. *International Journal of Scholarly Research in Engineering and Technology*, *4*(1), 1–18.

Spring, M., Faulconbridge, J., & Sarwar, A. (2022). How information technology automates and augments processes: Insights from Artificial-Intelligence-based systems in professional service operations. *Journal of Operations Management*, *68*(6–7), 592–618.

Tanna, M., & Dunning, W. (2023). Bias and discrimination in the use of AI in the financial sector. In *Artificial intelligence in finance* (pp. 320–349). Edward Elgar Publishing.

Torres, P. S., & George, B. (2023). Disruptive transformation in the transport industry: Autonomous vehicles and transportation-as-a-service. *International Journal of Emerging Trends in Social Sciences*, *14*, 28–37.

Tshabalala, L. (2023). The role of artificial intelligence in enhancing workforce productivity: Advanced computing applications in HR. *Journal of Advanced Computing Systems*, *3*(12), 26–33.

Vicente, L., & Matute, H. (2023). Humans inherit artificial intelligence biases. *Scientific Reports*, *13*(1), 15737.

Yanamala, K. K. R. (2024). Artificial Intelligence in talent development for proactive retention strategies. *Journal of Advanced Computing Systems*, *4*(8), 13–21.

Zhang, Q., & Xiong, Y. (2024). Harnessing AI potential in e-commerce: Improving user engagement and sales through deep learning-based product recommendations. *Current Psychology*, *43*(38), 30379–30401.

Zhou, W., Zhuang, Y., & Chen, Y. (2024). How does artificial intelligence affect pollutant emissions by improving energy efficiency and developing green technology? *Energy Economics*, *131*, 107355.

3 The Importance of the Shingo Model to the Human Side of Lean

THE LEGACY OF DR. SHIGEO SHINGO

Dr. Shigeo Shingo left a legacy in Lean manufacturing that few individuals can challenge. He was a key contributor to the development of total quality management (TQM), JIT manufacturing, SMED, poka-yoke (mistake proofing), and the concept of going to the gemba (actual workplace). Dr. Shingo authored 18 books in Japanese discussing these and other topics, seven of which have been translated into English. He was a consultant and coach to Toyota executives such as Taiichi Ohno, who championed the implementation of Dr. Shingo's concepts at Toyota (Shingo, 1986).

In 1988, Utah State University awarded Dr. Shingo an honorary doctorate in management. In the same year, the prestigious Shingo Prize was established based on Dr. Shingo's teachings and incorporated into the University. The Shingo Model™ provides a powerful framework designed to guide the transformation of an organization's culture toward the pursuit of excellence.

IMPORTANCE OF THE SHINGO FRAMEWORK

The quote from Dr. W. Edwards Deming best explains the basis of the Shingo model. He said, "It is not necessary to change. Survival is not mandatory" (Graban, 2017). This tongue-in-cheek quote by the quality guru educates us that change is a prerequisite for survival and growth.

Any organization that is not focused on continuous improvement and growth may not survive for long. Survival requires organizations to adopt a model that allows them to stay ahead of their competitors and their past achievements and accolades. Survival also requires a culture where every individual within the organization demonstrates humility, engagement, and empowerment and is focused on pursuing ideal results. The Shingo Model is a comprehensive framework to guide the transformation of an organization's culture toward achieving the goals of sustained organizational excellence.

The Shingo Model represents the relationships and interactions between tools, systems, principles, culture, and results. It provides a foundation for organizations to build current initiatives that take them forward on the journey to a sustainable culture of organizational excellence.

The critical requirement of the Shingo cultural transformation process is to align every organization, department, team, executive, manager, employee, tool, and system with the principles of sustainable operational excellence. When these are

DOI: 10.1201/9781003532972-3

appropriately aligned with Shingo principles, they strategically influence people's behavior and pull the organization toward an ideal culture.

The Shingo Model consists of five critical components intended to create a culture of organizational excellence. Also embedded within the Shingo Model are three fundamental paradigm shifts that enable the transformation of organizations. The transformation can be significantly accelerated when these paradigms become an integral part of all employees' corporate culture and daily behavior patterns. The three paradigm shifts are:

1. Organizational systems must be aligned with the appropriate principles and values to drive the intended behavior.
2. A direct and strong correlation exists between principles, systems, and tools.
3. A culture of operational excellence requires a focus on both behaviors and results.

(GBMP, n.d.)

FIVE KEY COMPONENTS OF THE SHINGO MODEL

The Shingo Model, shown in Figure 3.1, is represented by a diamond and has the following five elements: Guiding Principles, Systems, Tools, Results, and Culture.

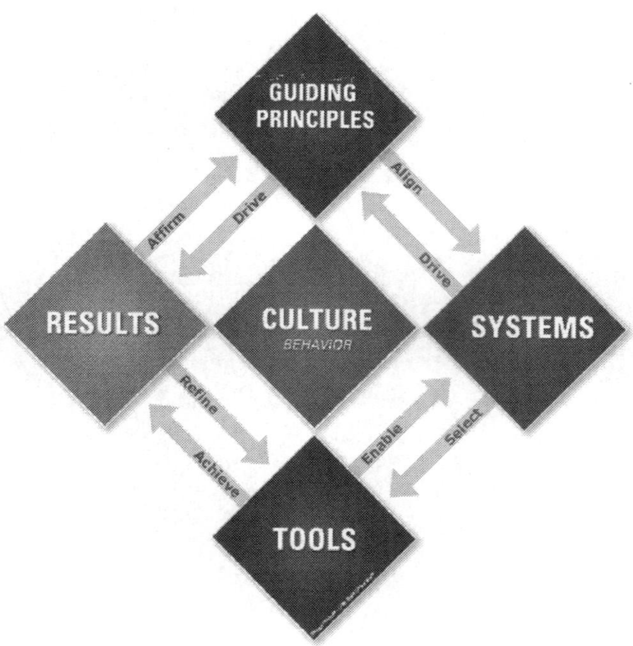

FIGURE 3.1 The Shingo Model.

GUIDING PRINCIPLES

A principle is a timeless and core axiom that governs the consequences of actions and behaviors. Shingo's guiding principles operate in the background, and when the actions or behaviors align with the principles, the consequences are generally good. The consequences are not good when actions do not align with the guiding principles.

The Shingo Model consists of ten guiding principles that lay the foundation of a culture of sustained organizational excellence. These ten principles are divided into three dimensions: 1. cultural enablers, 2. continuous improvement, and 3. enterprise alignment (Value Capture, n.d.) as shown in Figure 3.2.

The Shingo Model-related principles are universal and applicable across various industries, organizations, geographies, and individuals. These principles are based on creating a culture that encourages and requires asking the "why," which can ultimately lead to sustainable organizational excellence.

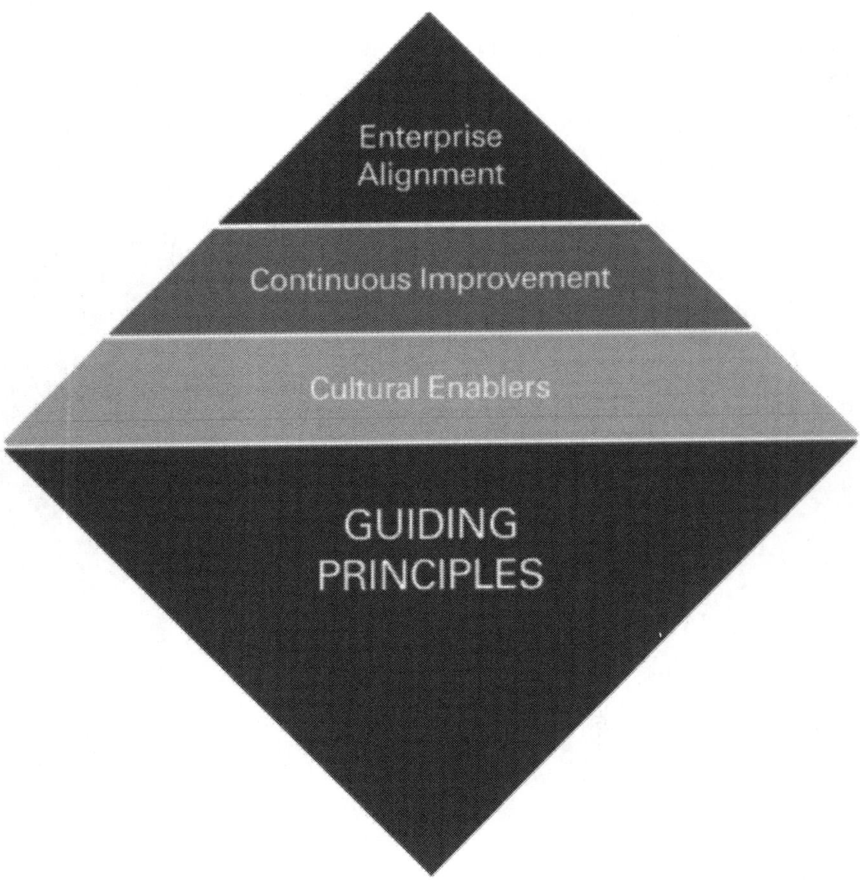

FIGURE 3.2 The guiding principles.

Systems: A system is a collection of tools, principles, or procedures working together as parts of a mechanism or an interconnecting network to accomplish an intended outcome. All organizations need to work as systems, not silos.

Tools: A tool is an object that can extend an individual's ability to modify features of the surrounding environment or help them accomplish a specific task.

Result: A measurable consequence, effect, or outcome, either successful or unsuccessful, results from implementing tools and systems – a consequence, effect, or outcome of something.

Culture: Culture encompasses all behaviors, customs, norms, and achievements of a group department, or organization.

BUILDING BLOCKS OF THE SHINGO MODEL

Tools, processes/systems, and principles drive the building blocks of the Shingo Model. The tools and systems reflect the "know-how," and the principles reflect the "know-why" within the organization. Organizational leaders, managers, and employees have varying responsibilities in implementing the Shingo Model, and they influence the culture and results based on their respective roles and behaviors. This is represented in Figure 3.3.

The Shingo Model was created to reflect guiding principles in the culture and behaviors that will ultimately impact the results. The model recognizes that the tools and systems are inadequate to operate a business. The core element of a company is always the people. Each person, team, and department within an organization has a

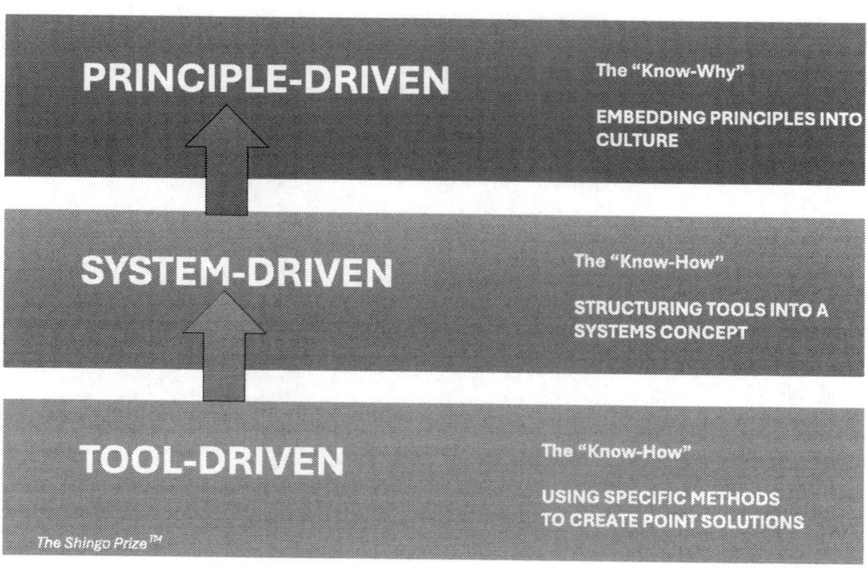

FIGURE 3.3 The building blocks of the Shingo Model.

set of values and beliefs that influence how they behave. The aggregate of behaviors is the organizational culture, which dramatically influences operational excellence and the organization's results. The results need to translate into value for the customers.

A significant challenge in most organizations is translating operational excellence into customer value as seen in Figure 3.4. If bridged correctly, this gap can lead to superior business results and profitability. The Shingo Model is the bridge that connects the organization's culture of excellence with the mission of creating customer value, as shown in Figure 3.5.

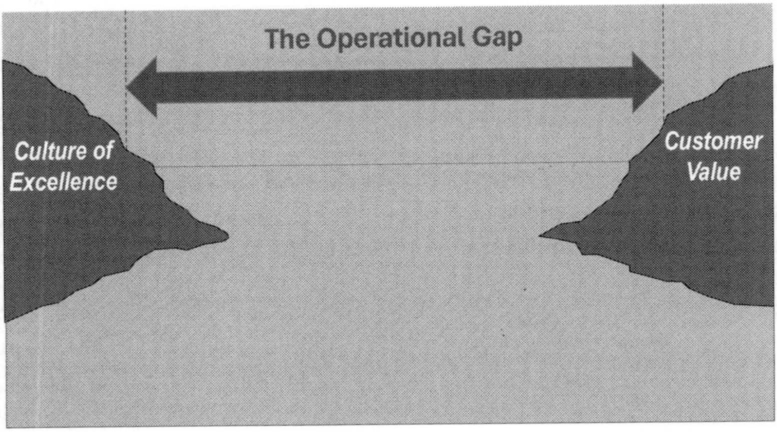

FIGURE 3.4 The operational gap in organizations.

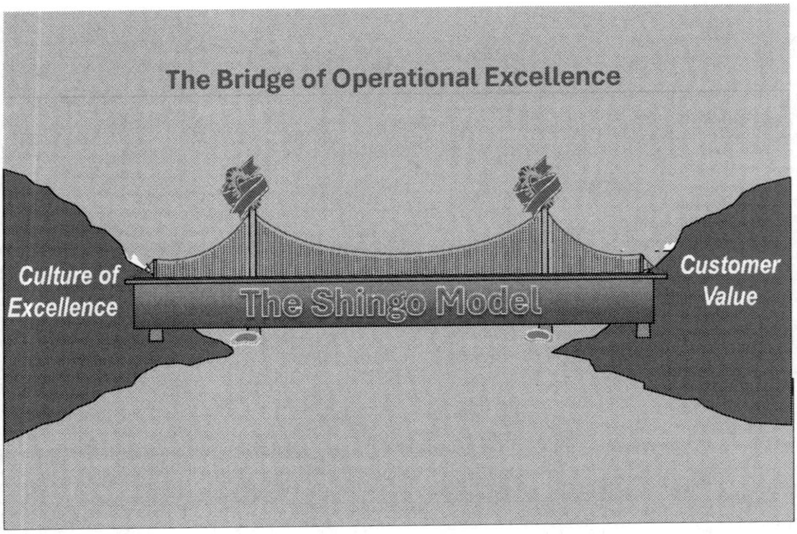

FIGURE 3.5 The Shingo Model Bridge.

THE TEN GUIDING PRINCIPLES OF THE SHINGO MODEL

The guiding principles are divided into three dimensions, namely cultural enablers, continuous improvement, and enterprise alignment. The ten guiding principles are cultural enablers, which include respect for every individual, and leading with humility; continuous improvement, which includes ensuring quality at the source, improving flow and pull, seeking perfection, ensuring scientific thinking, and focusing on the processes; and enterprise alignment, which includes creating value for the customer, creating a constancy of purpose, and thinking systematically, as shown in Figure 3.6.

In brief, the ten primary guidance provided by each of the ten principles are provided in Table 3.1.

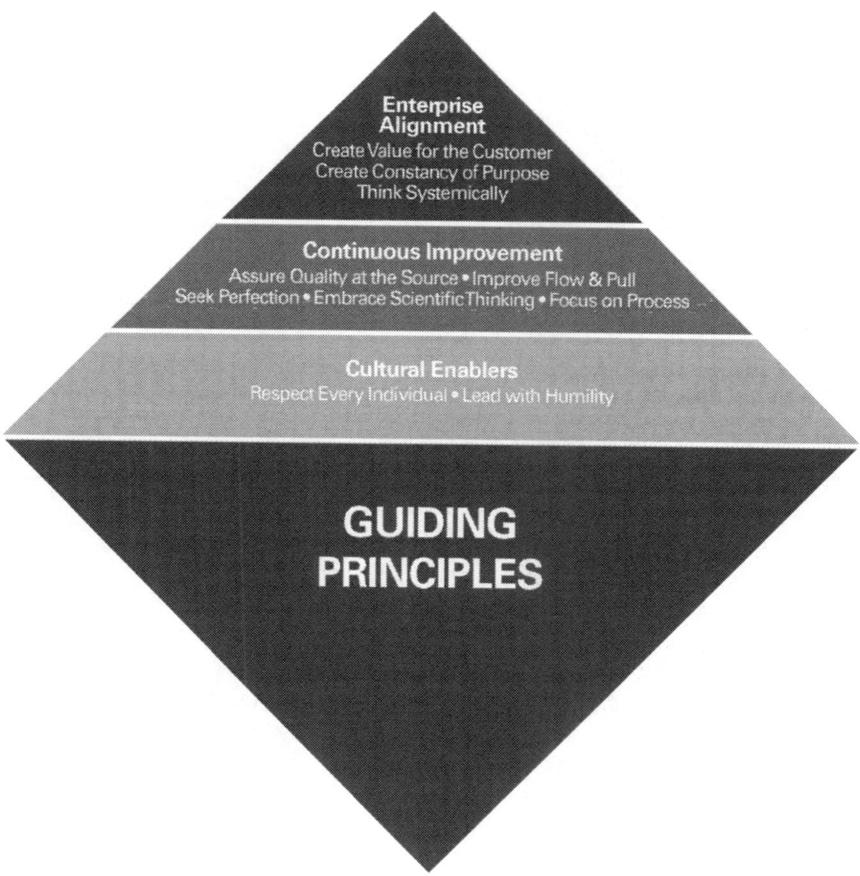

FIGURE 3.6 Shingo Model's ten guiding principles.

TABLE 3.1
Primary Guidance Provided by Each of the Ten Principles of the Shingo Model

#	Principle	Details
1	Respect every individual	This principle emphasizes the value of people in an organization, encouraging leaders to engage, empower, and nurture the development of their teams.
2	Lead with humility	Leaders must be open to feedback, willing to learn, and acknowledge that continuous improvement is a collaborative effort.
3	Seek perfection	Continuous improvement is the goal, aiming for excellence in all aspects of the organization while recognizing that perfection is a journey, not a destination.
4	Embrace scientific thinking	Decisions and problem-solving should be based on systematic, data-driven approaches and experimentation to foster learning and innovation.
5	Focus on process	Results are achieved through well-defined processes. This principle encourages organizations to prioritize improving processes to deliver consistent and high-quality outcomes.
6	Assure quality at the source	Build quality into the process by preventing defects from occurring, rather than relying on detection and correction afterward. This includes practices like poka-yoke and source inspection.
7	Flow & pull value	Ensure that the production or service flow is smooth and that work is pulled by demand rather than pushed by forecasts, reducing waste and improving efficiency.
8	Think systemically	Understand and optimize the entire system, recognizing that processes and functions are interconnected, and decisions should consider the broader impact on the organization.
9	Create constancy of purpose	Maintain a clear, consistent focus on long-term goals and values, ensuring alignment throughout the organization and with external stakeholders.
10	Create value for the customer	All activities and improvements should ultimately aim to deliver value to the customer, as they are the primary reason for the organization's existence.

These principles serve as the foundation for building a culture of operational excellence and are essential for sustaining long-term success.

As seen from the guiding principles, cultural enables are critical to directing the behavior of all employees. The best organizational cultures are built around ideal behaviors of humility, respect, trust, collaboration, innovation, and empowerment. Thus, these are the ideal behaviors that drive cultural transformations.

THE CRITICAL HUMANISTIC PRINCIPLES OF THE SHINGO MODEL

The two most important Shingo Model guiding principles that focus on the humanistic elements are respect and humility as shown in Figure 3.7.

FIGURE 3.7 The two humanistic pillars of Lean.

Most Lean initiatives fail due to an inability to adhere to a critical yet fundamental principle, "Lean is about people, not techniques" (Dalal, 2010). People are the primary appreciating assets in any organization. Some companies use this notion as a slogan, but only a few Lean organizations genuinely believe it. In most cases, initiatives' success (or failure) depends on the people executing them rather than on any tool, technique, equipment, consultant, or software.

Respect is the DNA of all strong cultures. Respect allows one to provide candid inputs based on their specific role, qualification, experience, and expertise. Respect means that the opinions and role of the individual are acknowledged, and the team members and the organizations welcome their contribution to the overall effort. Respect also means acknowledging the worth and potential of each individual, irrespective of the title. It also involves training, creating development plans, empowering, and mentoring or coaching them to enhance skills and performance (Loo, 2022).

One of the powers of great leaders is humility (Dalal, 2012). Humility does not mean thinking less of oneself but just thinking of oneself less. Organizations must encourage humility in leaders, which involves actively listening to others' opinions, accepting that others may have better ideas, and owning up to any mistakes. Humble leadership fosters collaboration and growth.

The human side of Lean technology stands on these two pillars of respect and humility. Without these two pillars, Lean technology would lack the focus on the human and cultural elements, leaving it just a set of tools and techniques versus a living philosophy.

In this book, we will use the two pillars of respect and humility to demonstrate that although AI will significantly enhance Lean in all areas, it can never replace the importance of humanity.

THE SHINGO MODEL IN THE AGE OF AI

Will the humanistic principles of the Shingo Model need to be updated in the age of AI? The answer is "Yes." In the age of AI, the Shingo Model will need to adapt to address emerging ethical and privacy concerns. The principles of respect every individual and lead with humility will need to incorporate safeguards for data privacy, ensuring that AI-driven systems do not violate the rights of employees or customers.

FIGURE 3.8 The four humanistic pillars of Lean in the age of AI.

Additionally, the ethical use of AI promotes transparency and fairness in data-driven decision-making must also be included. As AI becomes more integrated into business processes, the focus on creating value for the customer will also need to balance innovation with the ethical responsibility to protect personal data and privacy rights. Integrating these considerations will ensure that the model remains aligned with ethical leadership in a technology-driven world. In the age of AI, we will need to include four pillars of humanistic technology, including ethics and privacy as seen in Figure 3.8.

CONCLUSION

The Shingo Model, named after Shigeo Shingo, lays the foundation for the Shingo Prize established by the Utah State University in 1988. It is a standard that most organizations can follow to reach a level of excellence within their respective industry. This well-thought-out model consists of guiding principles, tools, systems, results, and culture as the building blocks for transforming organizational culture, which can result in profitability and customer value.

Respect and humility are the two principles that result in Lean, a philosophy that results in the growth of individuals and lays the foundation for transforming organizational cultures. AI can enhance the tools and techniques of Lean. However, it cannot duplicate these two critical elements of humanity, at least for now. In addition to the two humanistic pillars of respect and humility, the use of AI will necessitate the addition of two pillars of ethics and privacy.

REFERENCES

Dalal, A. (2010, September). Keep it simple. *Quality Progress, ASQ*, 25–28. https://www.linkedin.com/pulse/importance-respect-workplace-sai-choong-loo-fcpa/

Dalal, A. (2012, October). *The 12 pillars of project excellence: A lean approach to improving project results*. CRC Press.

GBMP. (n.d.). Shingo Institute courses. *GBMP*. https://www.gbmp.org/shingo-institute-courses

Graban, M. (2017). *The conundrum that is Dr. Deming on metrics, measures, and data*. https://www.leanblog.org/2017/07/conundrum-dr-deming-metrics-measures-data/

Loo, S. C. (2022). The importance of respect in the workplace. *LinkedIn*. https://www.linkedin.com/pulse/importance-respect-workplace-sai-choong-loo-fcpa

Shingo, S. (1986). *Zero quality control: Source inspection and the poka-yoke system*. Productivity Press.

Value Capture. (n.d.). Guiding principles of the Shingo Model. *Value Capture*. https://www.valuecapturellc.com/guiding-principles-shingo-model

4 Integrating Lean Tools with AI

TRADITIONAL VS. AI-ENHANCED LEAN TOOLS

Integrating AI into Lean tools significantly shifts from traditional methods, introducing automation, predictive capabilities, and real-time data analytics. Traditional Lean practices often rely on manual inspections, human decision-making, and retrospective analysis, which can be time-consuming, error-prone, and limited in scope. In contrast, AI-enhanced Lean tools leverage advanced technologies like machine learning, sensors, and predictive analytics to optimize processes in real time, increase accuracy, and reduce manual labor. The differences between traditional and AI-enhanced approaches across several Lean tools highlight the potential for substantial improvements in operational efficiency and quality (Antony et al., 2023).

In the case of 5S, workplace organization and cleanliness are traditionally maintained through manual checks, often requiring employees or supervisors to inspect areas to ensure tools and materials are in their proper places (Cudney et al., 2013). This manual approach can be inefficient, as human error or oversight may lead to missed issues. In an AI-enhanced system, computer vision or AI-powered sensors can automate this process, continuously scanning the work environment for disorganization, clutter, or potential safety hazards (Shahin et al., 2024). These technologies can immediately detect issues, providing real-time alerts and allowing for quicker responses. This shift leads to higher accuracy in maintaining standards and reduces the time spent on routine inspections, enabling employees to focus on more value-added activities. Quantifiable improvements from AI-enhanced 5S include faster detection of issues, increased adherence to organization standards, and reduced time spent on inspections.

For SMED, reducing changeover time relies heavily on employees' experience and manual adjustments. Changeovers are often initiated based on production schedules, with no predictive capability to foresee when they should optimally occur. AI-enhanced SMED, however, uses machine learning algorithms and real-time production data to predict the most efficient timing for changeovers (Söylemez & Ünlü, 2024). Organizations can use AI to analyze production output, equipment performance, and demand patterns to automatically suggest when changeovers should occur, optimizing the timing and the setup process. This results in reduced changeover times, fewer production delays, and improved machine utilization, significantly increasing flexibility and responsiveness.

Similarly, TPM traditionally involves scheduled maintenance activities, which are often based on fixed intervals rather than the machinery's actual condition. Scheduled maintenance can lead to either over-maintenance (wasting time and resources) or under-maintenance (resulting in unexpected breakdowns). With AI-enhanced TPM, sensors and real-time data analytics make predictive maintenance possible. These

DOI: 10.1201/9781003532972-4

systems continuously monitor the health of equipment, detecting subtle changes that could indicate future failures. Organizations can use AI algorithms to analyze historical performance data to predict when a machine will likely fail, allowing maintenance to be scheduled just before issues arise, thus preventing downtime (Antosz et al., 2020). This results in lower maintenance costs, reduced unplanned downtime, and extended equipment life. Organizations can significantly improve OEE by replacing reactive maintenance with predictive capabilities (Powell, 2024).

In VSM, traditional methods involve manual data collection and process mapping, often requiring considerable time and effort from employees to document and analyze the flow of materials and information through production processes. This approach is prone to data collection delays and analysis errors, making it challenging to implement timely improvements. AI-enhanced VSM automates data collection through IoT devices and sensors, providing real-time visibility into production flows. AI can quickly identify inefficiencies, bottlenecks, and waste, allowing organizations to make faster, data-driven decisions (Endrigo Sordan et al., 2024). The result is reduced lead times, improved process flows, and quicker waste identification.

Regarding kanban systems for managing inventory, traditional approaches rely on visual cards or signals to replenish materials. However, these systems can be prone to delays and errors, particularly when demand fluctuations are not immediately accounted for. AI-enhanced kanban systems use demand forecasting and real-time data to automatically adjust inventory levels based on actual consumption and anticipated demand. AI can predict when stock will run low and automatically trigger orders to replenish materials, minimizing the risk of overproduction or stockouts (Markham et al., 2000). This results in optimized inventory levels, reduced excess inventory, and improved material flow, leading to a more responsive and efficient supply chain.

In these examples, AI-enhanced Lean tools deliver quantifiable improvements in process efficiency, accuracy, and responsiveness. Reduced downtime, faster process adjustments, higher accuracy in inspections, and better resource utilization are just a few of the measurable benefits that organizations can realize through the integration of AI. By automating manual tasks, enhancing decision-making with data-driven insights, and enabling predictive capabilities, AI dramatically improves the potential of traditional Lean tools, allowing organizations to achieve higher levels of operational excellence and continuous improvement.

GAPS IN AI AND LEAN

While AI brings significant enhancements to Lean practices, there are important gaps that must be addressed to ensure successful integration. Similarly, Lean, though highly effective in optimizing processes, also has its limitations. Understanding these gaps is essential for creating a balanced and effective Lean-AI integration strategy.

GAPS IN LEAN

Lean is a powerful methodology, but it also has gaps that limit its potential in certain areas. A primary gap is the human factor. Lean often relies on human expertise and decision-making, which can introduce inconsistency. Human error, fatigue, or resistance to change can undermine Lean initiatives, particularly in areas requiring high precision.

Another challenge is the lack of consistency in training. Leadership buy-in and understanding of Lean principles can vary significantly across different levels of an organization. Without consistent training and leadership alignment, Lean initiatives often fail to gain traction or deliver sustainable results.

The issue of defining the goal is another gap in Lean. Different departments, such as finance and operations, often have divergent priorities. For instance, finance might prioritize cost reduction, while operations may focus on efficiency or quality improvement (Agustiady & Cudney, 2022). These conflicting goals can derail Lean efforts if they are not properly aligned.

Finally, resource constraints can limit the effectiveness of Lean. In some industries or organizations, the lack of skilled workers, capital, or other necessary resources hinders Lean's focus on efficiency. Lean requires a certain level of training, tools, and time investment to see long-term results, which not all organizations can afford.

GAPS IN AI

AI has tremendous potential, but several critical gaps can hinder its effective application. One key challenge is the lack of sufficient training. AI systems require a high level of expertise, and many organizations struggle with the skills gap. Without proper training, employees and the AI systems may not function effectively, leading to suboptimal results. Inconsistent or inadequate training programs can slow adoption and create resistance to change.

Another significant gap is bias in AI models. AI systems are only as good as the data they are trained on, and if the data is biased, the AI's decisions will reflect that. This bias can skew outputs, particularly in areas such as hiring, performance reviews, or customer targeting, leading to inequities or missed opportunities. To address this, organizations need to invest in robust model governance frameworks to regularly audit and mitigate bias.

Data quality also presents a significant hurdle. AI relies on accurate, high-quality data to produce reliable insights. Inconsistent, incomplete, or poorly organized data can limit AI's effectiveness. This challenge can derail AI projects if not addressed early on. Organizations must prioritize data management practices to ensure AI works with the best possible information.

Additionally, privacy and security concerns loom large. AI systems that handle sensitive personal data – whether employee performance data or customer information – can expose organizations to risks of breaches or misuse if not carefully governed. Strong data protection policies must be in place to safeguard this information and maintain trust with stakeholders.

Finally, the speed of adoption can be a barrier. While AI promises faster, more efficient processes, converting traditional systems to AI-enhanced operations is often slow. Technical complexities, organizational resistance, and lack of a clear implementation roadmap can delay AI adoption. Organizations need to manage this transition thoughtfully to avoid disruptions.

There is also the temptation to treat AI as a "hammer for every nail" – applying it to every problem, even when it may not be the best solution. AI should be judiciously applied where it can provide the most value rather than forcing it into processes where traditional Lean approaches might suffice.

HOW LEAN CAN SOLVE AI PROBLEMS

Lean methodology, with its emphasis on process improvement and waste elimination, provides powerful solutions for many of the challenges AI faces. One of Lean's core principles is VSM, which is ideal for eliminating waste and identifying process bottlenecks. Applying this to AI, organizations can ensure that AI initiatives focus only on high-value tasks and avoid unnecessary complexity or over-engineering.

In terms of data quality, Lean's "go to the gemba" principle – direct observation of where work happens – ensures that data is collected accurately and reflects real-world conditions. This approach can dramatically improve the data quality used to train AI models, leading to more reliable and effective AI outcomes.

Lean tools like SWOT and PEST analyses can also help organizations transition from the current state to the future state with AI. These tools provide a structured framework for assessing risks, identifying opportunities, and managing the complexities of AI adoption.

Lastly, Lean can play a significant role in governance. Lean's structured processes for continuous improvement can be leveraged to establish governance frameworks for AI, ensuring that models are regularly reviewed, biases are identified, and improvements are made.

HOW AI CAN SOLVE LEAN PROBLEMS

AI, in turn, can address many of Lean's limitations. For example, one of Lean's most significant challenges is the human element – the potential for errors, fatigue, and inconsistency. AI can automate routine, repetitive tasks, ensuring consistency and reducing the chance for errors. By taking over tasks such as defect detection, inventory management, or equipment maintenance, AI allows human workers to focus on more strategic, creative tasks.

In terms of training, AI can provide consistent, automated learning modules that ensure every employee receives the same high-quality training. This consistency removes the variability often arising from traditional training programs and ensures employees have the necessary skills to implement Lean practices effectively.

AI can also help define organizational goals by analyzing data to suggest which goals to prioritize. AI systems can assess cost of quality (COQ) models and recommend where resources should be focused to drive the most impactful improvements.

Finally, AI can optimize Lean's focus on efficiency and resource management. AI systems can predict where defects are likely to occur and suggest preventive measures, effectively reducing waste and improving overall efficiency. This predictive capability can streamline resource allocation, ensuring materials, time, and labor are used as efficiently as possible.

AI-DRIVEN METRICS AND FEEDBACK SYSTEMS

AI has the potential to revolutionize how metrics are developed and used in Lean tools by providing real-time feedback systems that ensure continuous improvement and alignment with organizational goals. Traditional Lean practices rely on manually collected data and periodic reports to track performance, which can delay responses to inefficiencies and create gaps between data collection and actionable insights. With AI, organizations can transition to real-time data analytics that constantly monitor processes and instantaneously generate actionable insights, allowing quicker interventions and more accurate decision-making.

AI-powered systems can automatically gather vast amounts of data from sensors, IoT devices, and other production equipment to create dynamic dashboards that offer real-time visualizations of key performance metrics. Organizations can tailor these dashboards to display a range of critical metrics such as equipment utilization, production speed, defect rates, and process bottlenecks. By providing this information continuously, AI enables employees and managers to make informed decisions without waiting for periodic updates, ensuring that process adjustments can be made instantly, leading to greater operational efficiency.

One of the most powerful features of AI-driven metrics is their ability to predict future performance. Using machine learning models, AI can analyze historical data alongside real-time input to forecast potential issues before they arise. For instance, AI can predict equipment failures by detecting subtle patterns in machinery data that indicate wear and tear, enabling maintenance teams to address problems proactively. Similarly, AI can forecast production delays by identifying trends in production speed or demand fluctuations, allowing managers to adjust schedules, resources, or inventory levels to avoid disruptions. This predictive feedback system helps organizations prevent downtime, optimize production schedules, and reduce waste, all while ensuring that Lean principles of efficiency and waste reduction are upheld.

AI-driven feedback systems can also provide continuous improvement insights. For example, by constantly analyzing production data, AI can identify areas where performance deviates from expected standards or where efficiency can be improved. In a kanban system, AI could track inventory movements and notify managers when stock levels are consistently higher or lower than necessary, suggesting adjustments to prevent overproduction or shortages. In VSM, AI might highlight bottlenecks that occur repeatedly and recommend process changes to eliminate them. This granular, real-time feedback level empowers teams to take immediate corrective action, promoting a culture of ongoing refinement and enhancement.

In addition to predictive insights, AI-powered metrics help align daily operations with broader organizational goals by providing visibility into how specific processes contribute to overall performance. For instance, AI dashboards can display critical indicators such as cycle time, throughput, and first-time quality, linking these metrics directly to overarching business objectives like customer satisfaction, cost reduction, and delivery time (Buffon et al., 2024). When employees and managers have access to this real-time feedback, they can see the direct impact of their efforts on company goals, fostering greater accountability and engagement. This continuous alignment between operational performance and strategic objectives ensures that Lean practices are efficient and aligned with long-term business success.

Overall, AI-driven metrics and feedback systems give organizations unprecedented real-time ability to monitor, predict, and improve Lean processes. By offering dynamic dashboards, predictive models, and actionable insights, AI empowers teams to make more data-driven decisions that drive continuous improvement and operational excellence. Integrating AI into Lean tools ensures organizations can stay agile, responsive, and focused on long-term success, transforming traditional Lean metrics into powerful, future-focused feedback systems.

CHANGE MANAGEMENT AND CULTURAL CONSIDERATIONS

Integrating AI into Lean practices requires technological advancements and a significant organizational cultural shift. Successfully embedding AI into Lean tools demands a change in how employees view and interact with both processes and technology. For many organizations, introducing AI can be seen as disruptive, leading to resistance or fear of job displacement. Therefore, effective change management is crucial to ensure that AI is perceived as a tool that enhances human capability rather than replacing it. This cultural shift involves fostering collaboration between humans and AI systems, developing strategies to address resistance, and maintaining open communication to build trust in AI integration.

One of the most critical factors for successful Lean-AI integration is ensuring that employees are adequately trained to work with AI technologies. AI systems can provide valuable insights and automate repetitive tasks, but the ultimate success of AI in a Lean environment depends on effective collaboration between humans and machines. Employees must be trained on the technical aspects of AI tools and how to interpret AI-driven data, use predictive insights to make informed decisions, and leverage AI to optimize processes. This training should emphasize that AI is a collaborative partner, capable of augmenting human abilities by performing complex calculations, analyzing data, and identifying patterns more quickly and accurately than humans alone. By fostering this collaborative mindset, employees can feel empowered by AI rather than threatened by it, viewing AI as an essential tool that enhances their role in achieving Lean efficiency.

To overcome resistance to change, organizations must recognize that employees may feel uncertain or concerned about the impact of AI on their jobs. There is often a perception that AI will replace human employees, leading to job losses or diminished roles. To address these fears, leadership must adopt strategies that reinforce AI as an enhancement, not a replacement. One such strategy is to highlight AI's value in eliminating repetitive, low-value tasks, freeing employees to focus on more strategic, creative, and decision-making activities. For example, AI can automate data collection, error detection, and predictive maintenance, allowing employees to dedicate more time to problem-solving, process improvement, and innovation. By showing how AI supports human expertise, organizations can help shift the focus from job loss to job transformation, illustrating that AI can make work more fulfilling and less mundane.

Open communication and transparency are key to building trust in AI systems and ensuring employees feel confident implementing new technologies. Organizations should maintain clear and honest communication about how AI will be used, what data it will collect, and how that data will be utilized. It is crucial to involve employees in discussions about AI integration early on, soliciting their feedback and addressing their concerns openly. Transparency also means providing insights into how AI makes decisions and offering employees visibility into the data that drives AI's predictions and recommendations. When employees understand AI's role and can see the rationale behind its outputs, they are more likely to trust and embrace it. This trust-building process is essential for AI to be fully adopted within a Lean framework, as employees need to feel secure that AI is working in their best interest and aligned with the organization's goals.

Creating an organizational culture that encourages continuous learning is vital for sustaining successful Lean-AI integration. AI technologies and Lean practices will continue to evolve, and organizations must cultivate a culture where employees are open to learning and adapting to new tools and methodologies. Ongoing training, access to resources, and opportunities to experiment with AI applications will help employees remain engaged and capable of utilizing AI to its full potential.

Successful Lean-AI integration requires more than just implementing new technologies – it demands a cultural transformation. Organizations can create an environment where AI is seen as a valuable asset that complements and enhances human capabilities by prioritizing training and collaboration between humans and AI, addressing resistance to change, and ensuring open communication and transparency. This cultural shift will enable organizations to unlock the full potential of AI-driven Lean practices, resulting in greater efficiency, innovation, and long-term success.

LEAN-AI INTEGRATION APPROACH

Integrating AI with Lean tools allows organizations to automate routine tasks, enhance decision-making, and gain real-time data insights, making processes more efficient, accurate, and sustainable. AI can transform traditional Lean tools by introducing predictive and prescriptive capabilities, allowing for more proactive management and continuous improvement.

When integrating Lean tools with AI, the approach can be framed across three key aspects: operational, tactical, and strategic, as shown in Figure 4.1. We address

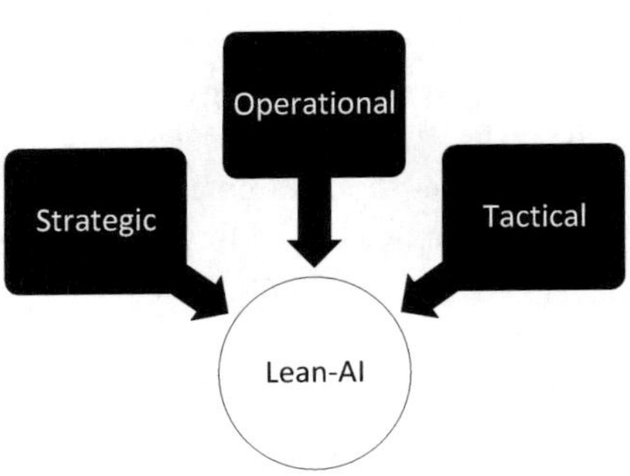

FIGURE 4.1 The three levels of application of Lean-AI.

integrating Lean tools with AI from tactical, operational, and strategic perspectives. Each level contributes to the comprehensive transformation of processes, ensuring immediate and long-term gains (Wilson et al., 2024). This tiered approach allows us to manage complexity systematically, align objectives with outcomes, and achieve sustainable improvements. Each level plays a vital role in successfully combining AI technologies with Lean methodologies, creating synergy between human expertise and technological advancements.

OPERATIONAL PERSPECTIVE: FOCUSING ON IMMEDIATE AND DAY-TO-DAY ACTIVITIES

The operational level addresses the immediate, hands-on tasks on the shop floor or within daily operations. These tasks, like equipment maintenance, changeovers, and workspace organization, are critical to keeping the organization functioning smoothly on a day-to-day basis.

At the operational level, the focus is on automating day-to-day activities and enhancing immediate processes. AI helps streamline routine tasks associated with Lean tools like 5S, SMED, and TPM. For example, AI can automate data collection during daily inspections, reducing human effort in sorting, cleaning, or maintaining machinery. Predictive maintenance and sensors in TPM can identify equipment issues before breakdowns occur, improving daily operations and reducing downtime (Agustiady & Cudney, 2024). Similarly, AI integration with SMED can automate changeover operations, reducing setup time by utilizing real-time data analysis to predict equipment needs. AI also enhances 6S (a version of 5S that adds safety), where predictive algorithms monitor and maintain safety standards and operational cleanliness in real time.

By addressing the operational perspective, we ensure that AI integration delivers immediate productivity benefits, reducing errors and enhancing consistency in daily operations. It helps to remove human error from mundane tasks and shifts employee focus to value-added activities.

TACTICAL PERSPECTIVE: ENSURING PROCESS EFFICIENCY AND CONSISTENCY

The tactical level focuses on optimizing the core processes that support the organization's overall functioning. This level is where we concentrate on improving workflows, ensuring standardization, and maintaining the consistency necessary for efficient operations.

At the tactical level, AI is crucial in optimizing processes and maintaining consistency across various Lean tools. AI's capability to analyze vast amounts of real-time data can be used to develop dynamic and adaptive maintenance schedules, such as in TPM, where it optimizes preventive and predictive maintenance activities based on historical data. In 6S, AI systems can ensure adherence to standardized processes by providing real-time feedback on sorting and organizing tools, thereby maintaining cleanliness and operational efficiency across multiple departments. In SMED,

AI refines operational workflows by analyzing bottlenecks in the setup process and suggesting real-time improvements for more effective transitions. AI-driven systems enable seamless integration of Lean practices into daily operations by monitoring performance metrics and improving decision-making capabilities.

At this level, AI is critical for maintaining process stability and consistency, ensuring that improvements made at the tactical level are sustained and enhanced through optimized processes. This focus creates a smooth, continuous workflow with fewer disruptions and higher efficiency.

STRATEGIC PERSPECTIVE: DRIVING LONG-TERM CONTINUOUS IMPROVEMENT

The strategic level focuses on long-term goals, such as enhancing competitiveness, achieving sustainability, and fostering continuous improvement. This perspective is vital because it ensures that AI integration aligns with the organization's broader goals and drives sustainable growth.

From a strategic perspective, integrating AI with Lean tools helps organizations pursue long-term continuous improvement and operational excellence goals. AI supports decision-making and sets strategic priorities by identifying trends and patterns that might not be evident through traditional methods. In the TPM context, AI aids in refining equipment maintenance strategies to extend machine life and reduce costs, aligning with an organization's broader goals of sustainability and productivity. AI's predictive analytics support strategic 6S initiatives by providing data-driven insights into workplace optimization, which contributes to the organization's sustainability and safety goals. Similarly, in SMED, AI allows for more strategic planning of production schedules by forecasting demand and ensuring equipment availability aligns with market needs.

Addressing AI integration from the strategic level ensures that the changes made at the tactical and operational levels are part of a long-term vision of sustainability and competitiveness. It allows organizations to solve immediate issues and position themselves for future growth and innovation.

By addressing Lean and AI integration across tactical, operational, and strategic perspectives, we ensure a comprehensive approach that tackles both short-term improvements and long-term sustainability. The tactical level delivers immediate productivity gains, the operational level ensures consistent and efficient processes, and the strategic level aligns AI integration with organizational goals for future growth and competitiveness. This tiered approach ensures that AI enhances Lean tools in a balanced, holistic way, driving continuous improvement at all levels of the organization.

IMPLEMENTATION PLAN

Integrating AI with Lean tools requires a structured, step-by-step approach to ensure the technology is adopted efficiently and delivers meaningful results. This high-level implementation plan includes four key phases: assessing readiness, selecting the appropriate technology, conducting pilot tests, and scaling and sustaining improvements. Each phase is critical for ensuring that AI integration is technically sound and aligns with the organization's operational goals and culture.

The first step in this process is assessing readiness, which involves evaluating which processes within the organization stand to benefit the most from AI. This assessment requires a thorough analysis of current operations to identify bottlenecks, inefficiencies, or areas where human error is a significant risk. Data-driven, repetitive, or complex decision-making processes are prime candidates for AI integration. For example, activities such as predictive maintenance, quality control, inventory management, and real-time scheduling may show the greatest potential for AI-driven optimization. During this readiness assessment, it is also important to consider the organization's current technological infrastructure, workforce capability, and culture to ensure the AI implementation is feasible and well-received. The goal is to prioritize processes where AI can provide the most immediate and significant value while also preparing the organization to handle the upcoming changes in workflow and culture.

Once the processes are identified, the next phase is technology selection. This step involves choosing the right AI tools and technologies based on the needs identified during the readiness assessment. Depending on the processes being optimized, organizations may opt for different AI technologies, such as machine learning algorithms for predictive analytics, computer vision systems for automated quality inspections, or IoT sensors for real-time data collection and equipment monitoring. Each of these AI technologies has its own strengths, and selecting the right one depends on the specific goals of the Lean process being enhanced. For example, a company focused on reducing machine downtime may benefit from predictive maintenance solutions that use machine learning, while a company looking to improve workspace organization could leverage computer vision to detect inefficiencies automatically. Additionally, organizations should consider the scalability, compatibility with existing systems, and ease of use of these technologies to ensure that they can be integrated smoothly and deliver sustained benefits.

The third phase involves pilot testing, where small-scale implementations of AI with Lean tools are deployed to test their effectiveness. Pilot testing allows organizations to validate AI's impact on specific processes in a controlled environment before scaling up. This phase helps identify potential challenges or bottlenecks in the integration process, such as data quality issues, technical limitations, or employee resistance to new technologies. During the pilot, teams should closely monitor key performance metrics – such as process speed, error rates, or equipment uptime – and compare these to baseline data collected before AI implementation. The goal of the pilot is to gather concrete evidence of AI's ability to improve the process while ensuring that any operational or cultural challenges are addressed early on. Additionally, feedback from employees involved in the pilot is crucial, as it can provide insights into how the AI system is being used in practice and whether further adjustments are needed.

Finally, once the pilot test demonstrates success, the next step is scaling and sustaining improvements. In this phase, the AI-enhanced Lean tools are rolled out across additional departments or throughout the entire organization. Scaling requires a systematic approach to ensure consistency, with well-defined procedures for integrating AI into new areas, training employees, and continuously monitoring performance metrics. It is important to expand gradually to ensure that teams have the necessary support and that any unforeseen issues are addressed as they arise. Sustaining the improvements achieved through AI integration involves continuously gathering data, refining AI models, and making iterative improvements to the system. This phase also requires ongoing employee training to ensure they remain proficient in using AI tools and continue engaging with Lean-AI practices as part of their daily workflows. Organizations should also establish governance and maintenance protocols to ensure that the AI systems are regularly updated, reliable, and aligned with evolving business objectives.

In summary, a high-level implementation plan for integrating AI with Lean tools involves a careful, phased approach: assessing readiness to identify the most suitable processes for AI integration, selecting the appropriate technology, conducting small-scale pilot tests to validate the impact of AI, and scaling the improvements across the organization while ensuring sustainability. This structured process ensures that AI-enhanced Lean tools are deployed to maximize their potential while minimizing disruptions and fostering long-term operational excellence.

ETHICAL AND PRIVACY CONSIDERATIONS IN AI-ENHANCED LEAN PRACTICES

Integrating AI into Lean practices offers significant opportunities for improving efficiency, decision-making, and waste reduction. However, it also introduces a range of ethical and privacy concerns that must be addressed to align with Lean's foundational principles of continuous improvement, respect for people, and waste elimination. While AI can enhance Lean tools such as VSM, 5S, and kanban, careful consideration is required to ensure that its use supports, rather than undermines, the human-centric philosophy at the core of Lean. Ethical issues related to employee privacy, data governance, and decision-making authority must be thoughtfully managed to maintain trust and ensure that AI is an enhancement, not a replacement, for human expertise.

One of the primary ethical concerns associated with AI integration is employee privacy, particularly when AI tools are used to collect and analyze data related to employee performance or behaviors. For instance, in VSM, AI can automate data collection from various parts of the production process to provide real-time insights into bottlenecks and inefficiencies. However, when this data collection involves tracking employees' actions or movements, it can raise concerns about surveillance. Employees may feel they are being constantly monitored, leading to a loss of trust, decreased morale, and potential resistance to AI systems. To address this, organizations must ensure that data governance policies are in place that define clear boundaries for what data is collected, how it is used, and who has access to it. Transparency is critical. Employees must understand how the data is used to improve processes, not penalize individual employees. Organizations can mitigate privacy concerns by maintaining open communication about data use and ensuring AI tools focus on process improvement rather than employee monitoring.

Data governance is another critical aspect that organizations must consider when integrating AI with Lean tools. AI systems rely on vast amounts of data to operate effectively, and the quality, security, and ethical handling of that data are paramount. Data security measures must be implemented to protect sensitive information from unauthorized access, especially when AI tools are used to analyze performance data that could include proprietary or personally identifiable information (PII). A robust governance framework should consist of data anonymization techniques, ensuring AI systems can optimize processes without compromising individual privacy. Additionally, organizations should establish clear guidelines for data retention and disposal, ensuring that data is only kept for as long as necessary to meet process improvement goals. Regular audits of AI systems and their data-handling practices should be conducted to ensure ongoing compliance with privacy standards and maintain employee and stakeholder trust.

Another essential consideration is employee autonomy and balancing AI-driven decision-making and human oversight. AI has the potential to provide powerful insights into process efficiency and can even suggest or automate certain decisions. For example, in a kanban system, AI might predict demand patterns and automatically adjust production schedules to prevent overproduction or stockouts. While this can improve efficiency, there is a risk that AI systems could be given too much control over decision-making, reducing employees' ability to exercise judgment or creativity. Lean philosophy emphasizes empowering employees to solve problems and contribute to continuous improvement, so AI must be implemented to support, rather than replace, human decision-making. Organizations should ensure that AI serves as a tool to enhance human judgment, offering data-driven recommendations while leaving final decisions to experienced employees. This balance maintains employee autonomy and allows AI to be seen as a supportive tool rather than a controlling force.

In tools like 5S, which focuses on workplace organization and cleanliness, AI can monitor adherence to standards through sensors and algorithms that track workspace conditions. For example, AI can detect when items are misplaced or areas become cluttered, providing real-time feedback to employees. While this can help maintain efficiency, ethical concerns about constant surveillance may arise if employees feel they are being watched too closely. To mitigate these concerns, organizations should involve employees in the integration process from the outset, explaining the benefits of AI for improving workplace organization and efficiency. By framing AI as a tool that helps employees maintain standards and avoid mistakes rather than a mechanism for surveillance, organizations can reduce resistance and foster a culture of collaboration between humans and AI.

Another critical ethical consideration is decision-making authority in AI-enhanced Lean practices. AI can provide advanced data analytics to identify patterns and make recommendations, but the ultimate decision-making authority should remain with humans. In processes such as root cause analysis or kaizen events,

AI can assist by analyzing large datasets to identify trends or potential causes of inefficiencies. However, the problem-solving process should still involve human creativity and collaboration. AI should be used to empower employees with better insights, not dictate actions or eliminate the need for human problem-solving. Ensuring that AI tools are considered advisory systems that enhance human decision-making helps maintain the core Lean principle of respect for people, fostering trust and cooperation in AI-integrated environments.

In summary, while AI can significantly enhance Lean tools like VSM, 5S, and kanban, it introduces ethical and privacy challenges that organizations must navigate thoughtfully. Data governance policies should be implemented to protect employee privacy and ensure the ethical use of data. Furthermore, AI should be positioned as a tool that supports human autonomy and decision-making, preserving the human-centric focus of Lean. By addressing these concerns – particularly around privacy, transparency, and employee involvement – AI can be successfully integrated into Lean practices to enhance efficiency and innovation while maintaining the trust and engagement of the workforce.

CONCLUSION

Integrating Lean tools with AI offers organizations a transformative pathway to enhance immediate operational efficiency and long-term strategic success. By combining AI across tactical, operational, and strategic levels, organizations can systematically optimize daily tasks, improve process consistency, and drive continuous improvement. At the tactical level, AI automates routine tasks, freeing up human resources to focus on more value-added activities, such as creative problem-solving and innovation. At the operational level, AI provides real-time insights and enables dynamic adjustments, ensuring that processes remain stable and efficient even in complex environments. Finally, at the strategic level, AI helps align Lean initiatives with broader organizational goals, such as sustainability, cost-efficiency, and responsiveness to market demands, fostering long-term growth and competitiveness.

While integrating AI with Lean practices holds immense potential, it also introduces ethical and privacy considerations that organizations must address thoughtfully. Issues such as data transparency, employee privacy, and maintaining human decision-making authority are critical to ensuring that AI enhances, rather than undermines, Lean's core principles of continuous improvement and respect for people. AI should be seen as a tool that empowers employees and supports their decision-making processes rather than replacing human judgment. Organizations can maintain trust by prioritizing transparency, involving employees in AI integration, and ensuring that AI complements human expertise while driving innovation and operational excellence.

Looking ahead, the future potential of Lean-AI integration extends far beyond current applications. Over the next five to ten years, we can expect significant advancements in how AI enhances Lean practices. One promising development is the emergence of AI-driven Lean consulting. AI systems act as virtual consultants capable of analyzing large datasets from an organization's operations and providing real-time recommendations for process improvement. These AI systems would autonomously identify inefficiencies, bottlenecks, and waste in production processes, suggesting tailored Lean improvements based on historical and real-time data. This capability could radically reduce the time and effort required for traditional Lean assessments while offering more precise, data-driven recommendations that adapt to real-time changes in the production environment.

Furthermore, the development of AI-powered autonomous Lean systems is on the horizon. These systems could continuously monitor all aspects of an organization's operations, autonomously implementing Lean adjustments without the need for human intervention. For example, AI systems could dynamically adjust production schedules based on real-time demand forecasts, automatically reconfigure workspaces to enhance flow and efficiency, or initiate maintenance activities based on predictive models. As AI continues to advance, these systems may also learn from past improvements and self-optimize over time, driving continuous improvement without the need for constant human oversight.

In addition, integrating machine learning and predictive analytics into Lean tools could lead to more predictive and prescriptive Lean systems. Instead of simply reacting to current inefficiencies, AI would predict potential future issues before they arise, allowing organizations to adjust processes and avoid problems preemptively. This integration would shift Lean practices from a reactive to a proactive approach, greatly enhancing the efficiency and resilience of organizations.

In terms of broader organizational impact, AI-driven Lean practices will likely contribute to sustainability initiatives, helping companies reduce energy consumption, minimize waste, and create more eco-friendly processes. AI's ability to analyze vast amounts of data will enable organizations to identify new opportunities for reducing resource use and improving environmental performance, further aligning Lean practices with global sustainability goals.

However, the importance of addressing ethical implications will grow as AI evolves. The rise of autonomous systems raises questions about accountability, decision-making authority, and potential unintended consequences. Organizations must establish clear governance frameworks to ensure that AI-driven systems align with their values, protect employee privacy, and maintain a human-centered approach to improvement.

In conclusion, integrating AI with Lean tools fundamentally enhances the efficiency and effectiveness of Lean methodologies, providing organizations with powerful new capabilities for optimizing processes and achieving operational excellence. As AI technology advances, its role in Lean practices will evolve, offering even greater potential for real-time, autonomous, and predictive improvements. However, organizations must also remain vigilant in addressing the ethical and privacy challenges that accompany AI, ensuring that Lean-AI integration remains true to the principles of continuous improvement and respect for people. By doing so, AI will enhance Lean practices and drive long-term innovation, sustainability, and competitiveness in an increasingly complex business landscape.

EXERCISES

1. Identify the tactical, operational, and strategic benefits of integrating AI into your current processes.

Process Area	Tactical Benefits	Operational Benefits	Strategic Benefits
Example: Maintenance Process	*Faster issue detection and reduced downtime*	*Optimized preventive maintenance scheduling*	*Extended equipment lifespan, lower long-term costs*

2. Analyze the current limitations in your processes and how AI can address them.

Process Area	Current Limitations	AI Approach to Overcome Limitations
Example: Inventory Management	*Inaccurate stock levels, human errors*	*AI-powered automated inventory tracking and optimization*

3. Develop a strategy for integrating AI into the tactical, operational, and strategic levels of your processes.

Process Area	Tactical AI Integration Strategy	Operational AI Integration Strategy	Strategic AI Integration Strategy
Example: Quality Control	*Implement AI-driven defect detection*	*Use AI for real-time process adjustments*	*Leverage AI for long-term quality trend analysis*

4. Evaluate the potential ethical challenges and data privacy concerns in implementing AI into your processes.

Ethical or Privacy Concerns	Impact on Workforce or Organization	Proposed Solutions
Example: Employee Surveillance	*Reduced morale due to constant monitoring*	*Implement transparency and clear data usage policies*

5. Reflect on the short- and long-term impacts of AI integration on your organization.

Area of Impact	Short-term Impact	Long-term Impact
Example: Productivity	*Immediate increase in efficiency*	*Sustainable growth in overall process performance*

6. Consider the technology requirements for AI integration and develop strategies for successful implementation.

Technology Required for AI	Integration Strategy
Example: Predictive maintenance	• *Integrate with existing equipment* • *Regular updates to AI algorithms*

7. Develop a high-level roadmap that outlines the steps necessary for successfully integrating AI with Lean tools. Include phases such as readiness assessment, technology selection, pilot testing, scaling, and sustaining improvements.

Phase	**Key Actions**
Example: Readiness assessment	*Identify processes that would benefit most from AI*
Technology Selection	
Pilot Testing	
Scaling	
Sustaining Improvements	

8. Identify specific areas within your Lean practices where AI could enhance operational efficiency, improve decision-making, or reduce waste. Consider AI applications in tools like 5S, kanban, or VSM.

Lean Practice	**Current Challenge**	**AI Opportunity**
Example: 5S	*Inconsistent workplace organization*	*Use AI-powered sensors to monitor and provide real-time alerts*

9. Reflect on the ethical challenges your organization may face when integrating AI with Lean tools. Consider how you will address concerns about employee privacy, decision-making autonomy, and data security. Propose strategies to handle these challenges responsibly.

Ethical Concern	**Potential Impact**	**Proposed Strategy**
Example: Employee privacy	*Risk of reduced employee trust due to AI monitoring*	• *Foster transparency* • *Involve employees in AI implementation decisions*

REFERENCES

Agustiady, T., & Cudney, E. (2022). *Building a sustainable Lean culture: An implementation guide*. New York, NY: CRC Press.

Agustiady, T., & Cudney, E. (2024). *Total productive maintenance: Strategies and implementation guide* (2nd ed.). New York, NY: CRC Press.

Antony, J., McDermott, O., Sony, M., Toner, A., Bhat, S., & Cudney, E. (2023). Benefits, challenges, critical success factors, and motivators of quality 4.0: A qualitative global study and directions for future research. *Total Quality Management & Business Excellence*, *34*(7–8), 827–846.

Antosz, K., Pasko, L., & Gola, A. (2020). The use of artificial intelligence methods to assess the effectiveness of lean maintenance concept implementation in manufacturing enterprises. *Applied Sciences*, *10*(21), 7922.

Buffon, C., Cudney, E., Elshennawy, A., & Furterer, S. (2024). Artificial intelligence techniques used for analyzing the voice of the customer in the quality 4.0 era: A systematic review using PRISMA protocol. *Journal of Educational Research*, *2*(3), 1–11.

Cudney, E., Furterer, S., & Dietrich, D. (2013). *Lean systems: Applications and case studies in manufacturing, service, and healthcare*. New York, NY: CRC Press.

Endrigo Sordan, J., Andersson, R., Antony, J., Lopes Pimenta, M., & Oprime, P. C. (2024). How industry 4.0, artificial intelligence and augmented reality can boost Digital Lean Six Sigma. *Total Quality Management & Business Excellence*, *35*(13–14), 1542–1566.

Markham, I. S., Mathieu, R. G., & Wray, B. A. (2000). Kanban setting through artificial intelligence: A comparative study of artificial neural networks and decision trees. *Integrated Manufacturing Systems*, *11*(4), 239–246.

Powell, D. J. (2024). Artificial intelligence in lean manufacturing: Digitalization with a human touch? *International Journal of Lean Six Sigma*, *15*(3), 719–729.

Shahin, M., Maghanaki, M., Hosseinzadeh, A., & Chen, F. F. (2024). Improving operations through a lean AI paradigm: A view to an AI-aided lean manufacturing via versatile convolutional neural network. *The International Journal of Advanced Manufacturing Technology*, *133*(11), 5343–5419.

Söylemez, İ., & Ünlü, R. (2024). Artificial intelligence in production systems and integration with lean manufacturing. In *Engineering applications of AI and Swarm intelligence* (pp. 235–252). Singapore: Springer Nature.

Wilson, R., Cudney, E., & Marley, R. (2024). Current status of Hoshin Kanri. *The TQM Journal*, *36*(2), 460–477.

5 6S-AI

5S OVERVIEW

The 5S framework is a Japanese methodology that helps factories improve efficiency and reduce waste using a simple "good housekeeping" approach. It is considered to be the foundation of Lean technology. Hiroyuki Hirano, a JIT expert, originally developed 5S (Hirano, 1990).

5S creates a clean, organized, and manageable work environment, which improves quality and productivity (Cudney et al., 2013). By providing a clean and organized work environment, 5S also creates a safer and more pleasant workplace.

5S is an acronym for the five Japanese words. Western organizations often translate these five Japanese terms into English derivatives also beginning with the letter "S". The five terms include:

Seiri: Sort
Seiton: Set in order or straighten
Seiso: Shine or scrub
Seiketsu: Standardize
Shitsuke: Sustain

As safety is critical to any Lean initiative, 5S has been updated to 6S. The 6S methodology consists of six steps. The steps to 6S progress toward creating a sustainable and clean environment. Therefore, the steps also progress from tactical to operational and then strategic (Dalal, 2010). These steps are translated from Japanese words starting with the letter "S" (Marchwinski & Shook, 2007).

STEP 1: SEFUTI AND SEKYURITI (SAFETY AND SECURITY)

Safety/Security is added as the first step of 6S, as it should be a priority for everyone in any organization to keep the workplace free of any possible injuries or danger to personnel or property. This step is a tactical, operational, and strategic step of 6S as it involves everyone in the organization.

STEP 2: SEIRI (SORT)

Sort is related to reducing the amount of occasionally used materials or tools that clutter the gemba (workplace). The sort step focuses on having only the necessary materials and tools for immediate production in the work area. This step is a tactical step of 6S, mostly involving the gemba personnel.

DOI: 10.1201/9781003532972-5

STEP 3: SEITON (SET IN ORDER OR STRAIGHTEN)

Set in order involves finding a fixed spot for all the sorted materials and tools for optimal efficiency. Set in order follows the age-old adage "a place for everything and everything in its place", which is accomplished by using outlines and shadow boards to facilitate creating a home for each item and designing a system that makes it easy to identify any missing objects. Set in order is a tactical step of 6S, mostly involving the gemba personnel.

STEP 4: SEISO (SHINE OR SCRUB)

Cleaning ensure the gemba areas and all items, including machinery, tools, floors, and walls, are always clean. It is important to note that the purpose of this step is not merely cleaning but cleaning with the intent to uncover any root causes of issues and to ensure timely preventative maintenance using visual cues. Shine is a tactical step of 6S, mostly involving the gemba personnel.

STEP 5: SEIKETSU (STANDARDIZE)

Standardization involves developing rules, processes, and procedures to ensure that 6S is applied consistently to all manufacturing, office, restrooms, and other areas. This step ensures the continuity and uniformity of achievements accrued by the safety, sort, set in order, and shine steps of 6S. Standardize is a tactical step of 6S, mostly involving managers and senior personnel.

STEP 6: SHITSUKE (SUSTAIN)

Sustaining improvements is one of the most critical steps of 6S and focuses on institutionalizing 6S using discipline, procedures, and metrics to maintain the gains achieved by implementing the 6S framework. This step can be achieved by using metrics, offering rewards, or using friendly departmental competitions to encourage the personnel to keep the focus and daily effort to maintain the 6S initiatives. Sustain is a strategic step of 6S as it mainly involves the executives.

ADVANTAGES OF 6S

6S is much more than just cleaning and housekeeping. 6S is the foundation for the Lean objectives of safety, quality, and efficiency. It is a systematic approach to creating and maintaining an organized, clean, high-performance workplace. 6S is also a conditioning discipline for any workplace (e.g., office, shop floor). It fosters a culture of accountability.

6S is a powerful tool that helps organizations with safety improvements, waste reduction, space-saving, machine uptime, and quality improvement. 6S helps create a safety-conscious workplace by reducing injuries and accidents by eliminating oil leaks, trip hazards, and other safety-related issues. 6S helps identify and eliminate waste or muda, such as transport, inventory, excess motion, waiting, overproduction, overprocessing, and defects. 6S helps free up space by removing unnecessary items and efficiently organizing necessary materials. Daily cleaning and addressing any abnormalities helps reduce machine breakdowns, thereby increasing machine uptime. 6S can help improve the quality of the work, which results in customer satisfaction. Mistakes are less likely to occur in a clean and organized environment (Liker, 2004).

LIMITATIONS OF 6S

For a 6S program to be successful, management must support all 6S efforts required for its long-term sustainment. Management must provide guidance, coordination, support, and unambiguous communications to create an environment where employees can employ their skills to the best of their abilities in a safe and well-organized environment (Agustiady & Cudney, 2023). Without appropriate support from management, understanding, or training of the true 6S model, the 6S initiative may be destined for failure from the start.

6S also relies on daily/weekly walk-throughs (gemba-walk) to identify potential hazards. Manual entries are also required daily for metric boards. 6S is a reactive model that relies on the personnel to identify potential issues and hazards. 6S initiatives are challenging to standardize and sustain as they require strict discipline and human intervention.

CULTURE

Since 6S is a human-centric process, during the 6S initiative, the focus on the two humanistic pillars of Lean – respect and humility is critical to the planning, execution, and sustainment of the 6S initiatives.

As with any methodology, there are limitations from a cultural perspective that must be considered. These limitations often stem from resistance to change, communication, differences in perceptions, entrenched work practices, and differing attitudes toward continuous improvement. There are also several limitations of 6S related to the cultural impact. The first limitation is the resistance to the changes proposed by the 6S methodology, which changes the traditional and established ways of working. This resistance often stems from preferring to maintain the status quo and a fear of inefficiencies being exposed to senior management.

Another key cultural limitation related to culture is the communication aspects of 6S. Due to the nature of 6S, open communication and feedback is necessary. For organizations that currently use indirect communication or where saving face is important, honest and direct feedback will be challenging. In addition, regular reporting of 6S progress and metrics is necessary for monitoring and sustainment. It can be seen as an additional responsibility by some and neglected, causing the 6S initiative to collapse if it becomes a systemic problem.

During the 6S event, differences in perceptions of cleanliness and order may cause conflicts among the team if communication is not open and direct. Individuals have varying perceptions of what constitutes cleanliness and order, which makes standardization challenging for organizations and could lead to conflicts or misunderstandings. 6S requires discipline and open communication to ensure gains are standardized and maintained. Discipline and communication styles vary across societies, generations, and individuals and can have an impact on 6S in multi-cultural and multi-national organizations.

Hierarchical structures can also cause reluctance to suggest improvements or take continuous improvement initiatives. Conflicts can arise based on the power structure between administrative offices and gemba, impacting the organization's sustainment and long-term commitment toward the 6S initiatives. Employees may also fear that their suggestions are seen as questioning authority or not communicating through the proper channels. 6S is a collaborative and participative Lean methodology that can clash with hierarchical dynamics.

Another primary cultural limitation of 6S is related to the routine nature of the 6S tasks and the human brain. The human brain is inclined toward efficiency by automating mundane, repetitive, and routine tasks in order to free up the "brainpower" for more challenging tasks. This ability of the brain can be detrimental in some cases, as humans tend to skip over important details or tasks, resulting in failures. For example, there is a very detailed step-by-step process that all aircraft pilots need to follow during take-offs and landings. However, since the human brain tends to skip over or minimize the importance of some important steps, pilots must follow a check sheet to prevent any steps from being missed. Since 6S requires several mundane and repetitive tasks to be followed, it is easy for some critical steps to be missed, leading to failures. As 6S is highly dependent on human intervention for its success, in the long term, this can lead to the failure of the overall 6S program.

INTEGRATING 6S AND AI

The success of 6S depends on human intervention and is a reactive model. Lean and AI, when used together, can transform 6S into a proactive framework. AI is a technology that enables the automation of tasks that typically require human intelligence. Therefore, the integration of 6S and AI is a pioneering effort that can help resolve the limitation of the requirement of regular human intervention during safety, sort, set-in-order, shine, standardize, and sustain steps of 6S. 6S-AI can make it easier to standardize and sustain 6S in the long term by automating mundane tasks and using the predictive capabilities of AI.

Organizations should foster a culture of continuous improvement and technological adoption to leverage the integration of 6S with AI fully. Encouraging collaboration between employees and AI systems, promoting openness to change, and providing training and support can help employees adapt to new technologies and processes. By embedding these cultural values, organizations can ensure AI's smooth and effective integration into 6S processes. The primary humanistic aspects of humility and respect for people can be enhanced as AI frees up people from mundane tasks and fosters collaboration on more collaborative and creative endeavors.

Integrating AI practices into 6S can address the limitations and further enhance the overall benefits of the methodology. First, rather than using 6S as a reactive approach, organizations can use existing data to predict and proactively address safety issues. When developing solutions to improve the work environment with 6S, we are limited to the experiences and ideas of our team. AI is a helpful approach to brainstorming and developing potential solutions. AI has the potential to be integrated within each of the six S's and significantly enhance the effectiveness of the overall 6S program.

Each 6S step can be enhanced through AI.

STEP 1: SEFUTI AND SEKYURITI (SAFETY AND SECURITY)

Organizations can use AI to analyze historical safety data and identify patterns of accidents or near misses that are not apparent in standard 6S weekly gemba walks. This analysis may involve investigating factors such as time of day, location, equipment, and environmental conditions. For example, AI might detect that a specific machine malfunctions at certain intervals, which leads to accidents. It also might determine that a common sequence of events precedes near misses.

STEP 2: SEIRI (SORT)

For the second S, organizations can use AI to help identify and categorize items in a work area. For example, machine learning algorithms can analyze usage patterns for supplies to determine if they are necessary and what quantities are appropriate. By eliminating unused items, the organization can declutter and optimize space. In addition, organizations can use AI to track and manage assets. AI can also be used to automate and streamline the sorting process in manufacturing and warehousing environments based on pre-defined criteria.

STEP 3: SEITON (SET IN ORDER OR STRAIGHTEN)

Organizations can employ AI to generate layouts and workplace organization recommendations through workflow analysis and ergonomic principles. The analysis can investigate the interaction between employees, tools, equipment, and materials. This information can be utilized to determine efficient people, equipment, and desk arrangements to minimize wasted movement and improve productivity.

STEP 4: SEISO (SHINE OR SCRUB)

AI sensors and monitoring systems can be used by organizations to detect or predict when equipment or areas need maintenance or cleaning. For example, predictive maintenance algorithms can predict when cleaning is necessary to ensure equipment runs in its best condition, reducing downtime. AI systems can use information on maintenance and cleaning schedules to determine the most efficient schedules.

STEP 5: SEIKETSU (STANDARDIZE)

Organizations can use AI to establish standard operating procedures (SOPs) to standardize cleaning processes and provide real-time information to employees. Tools such as augmented reality (AR) overlays and mobile applications can provide step-by-step instructions to ensure tasks are completed correctly and consistently each time.

STEP 6: SHITSUKE (SUSTAIN)

In the final S, sustain, AI can continuously monitor how well departments and work areas adhere to the 6S practices. Based on the level of adherence, AI can provide real-time feedback to employees to ensure consistency. In addition, organizations can use AI to help sustain the gains achieved through 6S by analyzing compliance level data and identifying improvement areas.

AI can also address one of the limitations regarding regular reporting of 6S progress and metrics by using real-time data and dashboards to ensure that the sustainment of 6S is not dependent on human intervention and can be automated. This is one of the primary advantages of 6S-AI that prevents the collapse of 6S due to a lack of human engagement.

INTEGRATING 6S WITHIN THE LEAN-AI MODEL

6S consists of tactical, operational, and strategic aspects for organizing and optimizing workspaces and processes, as shown in Figure 5.1. The tactical aspects of 6S-AI deal with the day-to-day activities. The tactical aspects of 6S-AI are the specific actions related to safety, sorting, setting in order, and shining, which aim to ensure workplace safety, organization, and cleanliness.

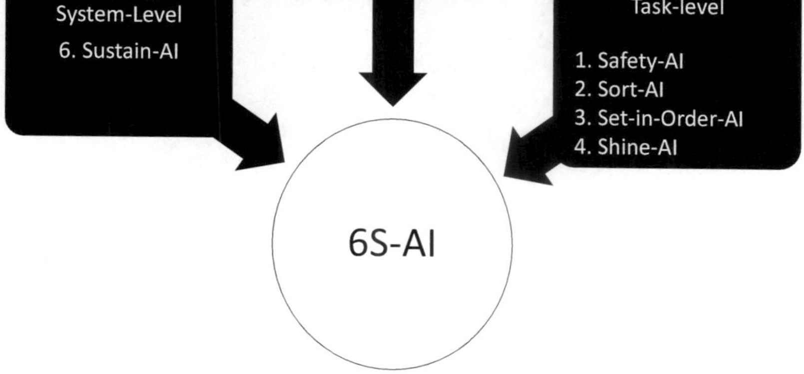

FIGURE 5.1 6S-AI model.

The operational aspects focus on the processes and procedures to ensure efficiency and consistency. Therefore, the sixth "S" of sustain is operational since it involves standardizing procedures and practices.

Finally, sustaining 6S-AI is a strategic aspect. The strategic aspect of 6S-AI is concerned with long-term planning and sustainability. The final "S", sustain, encompasses strategic planning and efforts to maintain the effectiveness of the 6S-AI program over the long term.

The integrated 6S-AI model provides specific tactical, operational, and strategic actions to ensure workplace safety, efficiency, and sustainability as seen in Figure 5.1.

6S-AI DATA PRIVACY AND ETHICS CONSIDERATIONS

When implementing 6S-AI, organizations should account for several critical data privacy and ethical considerations. The following elements ensure responsible AI use, especially in settings aiming to improve workplace organization and safety:

Data Collection Transparency and Consent: Clearly communicate to employees and stakeholders what data is being collected, why, and how it will be used. Obtain explicit consent, especially when collecting personal or sensitive data, to foster trust and respect for individual privacy.

Data Minimization and Security: Only gather data that is directly relevant to improving safety, efficiency, or productivity in alignment with 6S principles. Protect all collected data with robust cybersecurity measures to prevent unauthorized access, leaks, or misuse. Secure practices include encryption, access control, and regular audits.

Bias and Fairness in AI Algorithms: Ensure that the AI algorithms used in 6S applications, such as predictive maintenance or safety risk assessment, are free from bias and provide equitable outcomes. Regularly review AI outputs for disparities and have mechanisms in place for continuous improvement.

Data Anonymization/Coding: When feasible, anonymize data by coding or other means to protect individual identities, especially in environments where location tracking or employee performance data is collected. This can help reduce privacy risks and align with ethical principles, while still enabling AI to perform its intended functions.

Ethical Impact of AI Recommendations: Be cautious with AI-based decisions affecting employee roles or productivity, especially when these insights may impact job security, task allocation, or monitoring. Adopt a human-in-the-loop approach, where AI recommendations are reviewed by managers or employees before being fully implemented.

Transparency in AI Decision-Making: Design AI systems that provide understandable insights rather than opaque recommendations. For example, an AI system used to optimize workflows under the 6S model should clearly indicate why a process change is being suggested. This transparency allows for better employee buy-in and reduces feelings of alienation or resentment toward AI-driven changes.

Accountability and Governance: Implement a framework of accountability where AI-driven actions and outcomes are continuously monitored. Assign responsibility for overseeing data ethics, ensuring AI alignment with organizational values, and mitigating risks. Establish clear governance policies that define the ethical use of AI and empower employees to report privacy concerns without fear of reprisal.

Employee Empowerment and Engagement: Engage employees in the implementation process and solicit their feedback on how AI and 6S integration impacts their work environment. Address concerns, offer training, and provide support, as fostering a sense of shared responsibility is essential for the ethical adoption of AI and Lean methodologies.

In summary, organizations must adopt a balanced, ethical framework that ensures AI and 6S enhancements are privacy-conscious, transparent, and human-centered. This approach not only preserves trust but also amplifies the positive outcomes of Lean-AI integration for sustainable, people-first workplace excellence.

6S-AI INTEGRATION INDUSTRY EXAMPLES

The integration of 6S-AI is applicable across all industries. Organizations can use 6S-AI to implement innovative solutions to ensure safety, organization, cleanliness, and sustainability within their operations.

Manufacturing organizations can use AI-powered robots and automated guided vehicles to sort and organize materials in warehouses and production facilities. In addition, computer vision systems can identify and categorize items. Finally, AI algorithms can optimize inventory layouts and process flows for efficient access and movement of people, materials, and equipment.

Hospitals and healthcare facilities can employ AI-powered inventory management systems to maintain clean, well-organized supply rooms. These systems can forecast demand using predictive analytics to ensure critical medical supplies are always available when and where they are needed (JIT). In addition, AI-powered cleaning robots can ensure patient rooms and common areas are properly sanitized as part of the shine step of 6S.

In retail, AI-driven systems can optimize shelf space and product placement in stores. AI algorithms can analyze customer traffic patterns and purchasing behaviors to help retailers determine the most effective arrangement for product visibility and accessibility while reducing clutter. Retailers can also use AI-powered cleaning drones to maintain the cleanliness of large retail spaces efficiently.

From a supply chain perspective, organizations can use AI-driven warehouse management systems to help streamline logistics and warehousing operations. For example, the system can aid in organizing inventory, optimizing picking routes, and ensuring compliance with safety standards. In addition, the systems can continuously analyze data from sensors and IoT devices using AI algorithms to make real-time decisions that maintain orderliness and cleanliness.

In the food industry, organizations can use AI in processing plants to monitor equipment cleanliness and hygiene standards. Organizations can use sensors and AI algorithms to detect contaminants and anomalies in food production processes. These efforts ensure compliance with sanitation regulations and promote a clean working environment.

Finally, the integrated 6S-AI is also applicable in office environments. Organizations can use AI-powered smart office solutions to maintain a clean and organized workspace. For example, organizations can utilize sensors to monitor occupancy and usage patterns to optimize resource allocations and cleaning schedules. Organizations can use AI chatbots and virtual assistants to provide guidance and reminders for maintaining and sustaining the 6S principles.

6S-AI ADVANTAGES

Adding AI's predictive and generative capabilities to the 6S methodology accelerates the adoption rate of 6S and ensures long-term success. The predictive 6S-AI model is the future foundation of Lean-AI. 6S-AI includes several enhanced capabilities, such as data-driven decision-making, predictive maintenance, safety monitoring, workplace optimization, employee training and engagement, automated audits, collaborative robots, and smart facility management.

With respect to data-driven decision-making, AI can be employed to analyze data related to 6S activities, such as safety records, cleanliness metrics, and workflow efficiency. AI-driven dashboards can provide real-time insights into the status of 6S initiatives.

In terms of predictive maintenance, organizations can employ AI algorithms to predict maintenance needs for equipment and machinery, contributing to the sustainability aspect of 6S.

Organizations can monitor safety by integrating AI-based monitoring systems for workplace safety, such as using computer vision systems to detect unsafe behavior or potential hazards.

Workflow optimization can be achieved by using AI to analyze and optimize workflow processes, ensuring that the principles of 6S are applied efficiently.

Organizations can implement AI-driven training programs to educate employees on 6S principles, safety protocols, and best practices.

In addition, organizations can automate their audits by developing AI tools that can automatically audit processes for 6S compliance, reducing manual effort and ensuring consistency.

Organizations can integrate AI-powered collaborative robots (cobots) that can work alongside human employees to perform tasks, enhancing productivity and efficiency.

Finally, organizations utilize AI for smart facility management, including energy optimization, resource allocation, and predictive maintenance.

When integrating AI with 6S, it is essential to align technological solutions with the specific goals and principles of 6S methodology to enhance workplace efficiency, safety, and organization. Keep in mind that successfully integrating these elements requires a strategic and collaborative approach involving both technology and human factors.

CONCLUSION

6S is the foundation of any Lean initiative. 6S is a systematic approach that promotes a clean and organized workplace, which leads to enhanced safety, quality, and efficiency. The 6S model adds safety as the most critical part of this model. However, the major disadvantages of the current 6S model are that its success depends on human intervention and that it is a reactive model.

The future of Lean-AI requires a technology-driven foundation that we call 6S-AI. The advantages this AI-driven model will address are that it will not only address the human dependence issue but will make it predictive. 6S-AI helps organizations with data-driven decision-making, safety monitoring, workflow optimization, employee training and engagement, and facility management.

EXERCISES

1. What are the current system and process limitations of your 6S approach, and how can you apply AI to overcome these limitations?

6S Approach	System and Process Limitations	AI Approach to Overcome System and Process Limitations
1S: Sefuti and Sekyuriti (Safety and Security)		
2S: Seiri (Sort)		
3S: Seiton (Set in order or straighten)		
4S: Seiso (Shine or scrub)		
5S: Seiketsu (Standardize)		
6S: Shitsuke (Sustain)		

2. What are the current cultural limitations of your 6S approach, and how can you apply AI to overcome these limitations?

6S Approach	Cultural Limitations	AI Approach to Overcome Cultural Limitations
1S: Sefuti and Sekyuriti (Safety and Security)		
2S: Seiri (Sort)		
3S: Seiton (Set in order or straighten)		
4S: Seiso (Shine or scrub)		
5S: Seiketsu (Standardize)		
6S: Shitsuke (Sustain)		

3. What are some short- and long-term benefits to your department and organization for integrating AI into 6S?

#	Short-Term Benefits	Long-Term Benefits
Example	*Improved safety and security*	*Long-term cost savings*
1		
2		
3		
4		
5		
6		

4. Reflect on the ethical implications of using AI in 6S within your organization. Consider aspects such as employee privacy, job displacement, and data security. Develop strategies to mitigate the negative aspects of the workforce.

#	Ethical Considerations	Strategies to Mitigate
Example	*Reduced human interaction*	• *Foster human-AI collaboration* • *Maintain human oversight*
1		
2		
3		
4		
5		
6		

5. What are the technology considerations for integrating AI into your 6S framework within your organization? Develop strategies for the necessary technology integration.

#	Technology Necessary for AI	Strategy for Integration
Example	*Computer vision systems*	• *Integrate with existing surveillance* • *Regular system updates*
1		
2		
3		
4		
5		
6		

REFERENCES

Agustiady, T., & Cudney, E. (2023). *Building a sustainable lean culture*. New York, NY: Productivity Press.

Cudney, E., Furterer, S., & Dietrich, D. (2013). *Lean systems*. New York: NY: CRC Press.

Dalal, A. (2010). '6S' program: The most critical step on your lean journey. *ASQ Lean Enterprise Division Newsletter*, *9*(1), 2–4.

Hirano, H. (1990). *5 pillars of the visual workplace: The sourcebook for 5S implementation*. New York, NY: Productivity Press.

Liker, J. K. (2004). *The Toyota way: 14 management principles from the world's greatest manufacturer*. McGraw-Hill.

Marchwinski, C., & Shook, J. (2007). *Lean lexicon: A graphical glossary for lean thinkers* (4th ed.). Lean Enterprise Institute.

6 VSM-AI

VSM OVERVIEW

A VSM is the graphical documentation of all the actions (value-added, necessary non-value-added, and non-value-added) that occur to a product or service, beginning with its inception (at order) through shipment and delivery to the customer. It is a visual tool used in Lean methodology to map out, analyze, and design the flow of processes, materials, and information required to bring a product or service to a customer (Rother & Shook, 2003). It helps organizations visualize waste, streamline processes, and improve efficiency by providing a detailed overview of the entire production or service process from start to finish (Sharma & Sharma, 2020).

The name of the VSM is derived from three words:

Value: Indicates value-adding activities at every step of the stream

Stream: Refers to the complete flow of manufacturing or service operations (Silva & de Sousa, 2019).

Map: Indicates value stream activities at every step of the stream,

Note: Just like a regular map has a universally recognized language of symbols, a values stream also has a universally recognized graphical language that includes symbols and arrows.

KEY COMPONENTS OF A VALUE STREAM MAP

Customer Requirements: Information about what the customer expects, such as delivery time, quality, and specific product features (Melton, 2022).

Process Steps: All the steps involved in producing the product or service, from raw materials to finished goods and services delivered to the customer.

Material Flow: The movement of materials through the production process, often represented by arrows.

Information Flow: The flow of information that triggers and supports the material flow, such as orders, schedules, and instructions.

Timeline: A timeline at the bottom of the map showing the time taken for each step and the total cycle time, highlighting areas where delays or bottlenecks occur.

Value-Added and Non-Value-Added Activities: Differentiating between steps that add value to the product (value-added) and those that do not (non-value-added or waste or muda).

DOI: 10.1201/9781003532972-6

VSMs detail two types of scenarios:

Current State details the existing process

Future State details the possible scenario after removing the waste and implementing Lean tools (Hines & Rich, 1997).

VALUE STEAM MAPPING PHASES

At a high level, the VSM journey can be completed in three phases, as shown in Figure 6.1 (Duflou, & De Meulemeester, 2021):

1. Current state map
2. Action plan for improvement that includes kaikaku (radical transformation projects), and kaizen events.
3. Future state map

A DETAILED STEP-BY-STEP APPROACH TO VALUE STREAM MAPPING

Identify the Product or Service: Choose a specific high-value or high-importance product or service to map.

Map the Current State: Document the current process steps, material flow, and information flow from start to finish. This includes collecting data on cycle times, lead times, inventory levels, and more.

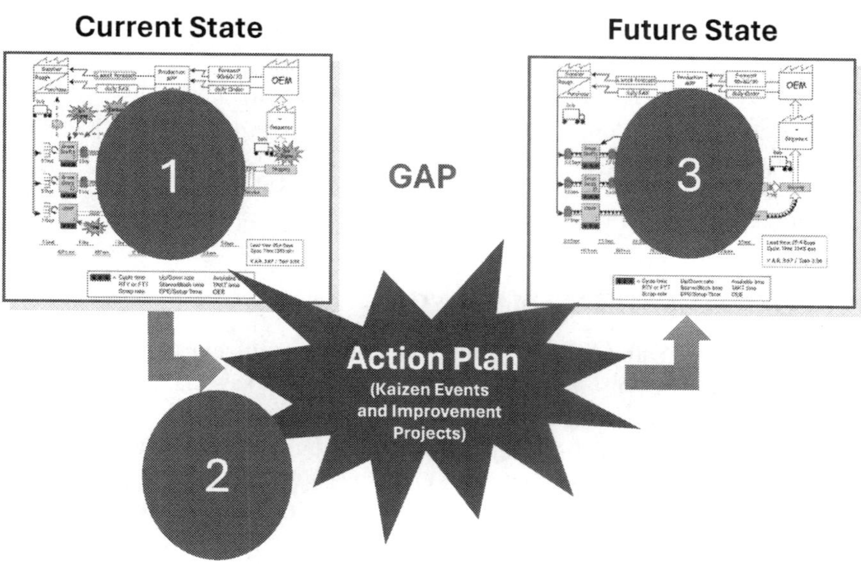

FIGURE 6.1 The three phases of value stream mapping.

Cell	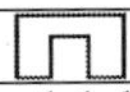	Operator	
Customer	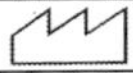	Pull	
Data	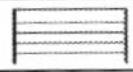	Push	
Electronic information		Raw materials	
FIFO		Shipment	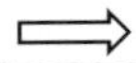
Go see		Signal kanban	
Inventory		Supermarket	
Kaizen burst		Truck	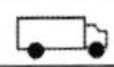
Kanban post		Verbal	
Load level		Withdrawal kanban	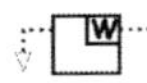
Manual information			
Database			

FIGURE 6.2 Value stream mapping symbols.

Analyze the Current State: Identify waste and inefficiencies in the current process. Look for delays, bottlenecks, excess inventory, unnecessary movements, and other non-value-added activities.

Design the Future State: Develop a vision for an improved process with reduced waste and increased efficiency. This involves brainstorming solutions to the issues identified in the current state analysis.

Implement Changes: Create an action plan to implement the improvements designed in the future state map. This may involve changes in workflows, equipment, training, or other resources.

Monitor and Adjust: After implementing changes, continuously monitor the process to ensure improvements are sustained and make further adjustments as needed.

Some VSM symbols (Six Sigma Material, n.d.) are seen in Figure 6.2.

TYPES OF VALUE STREAM MAPPING

Manufacturing Value Stream Map: Used to map manufacturing processes

Business Value Stream Map: Used to map business processes

A swim lane map (Rumbaugh & Blaha, 2019) is a better option for business VSM, as shown in Figure 6.3 (Rother & Shook, 2003).

ADVANTAGES OF USING A VALUE STREAM MAP

Overall Visibility: Provides a cross-function end-to-end visual representation of the entire product/service.

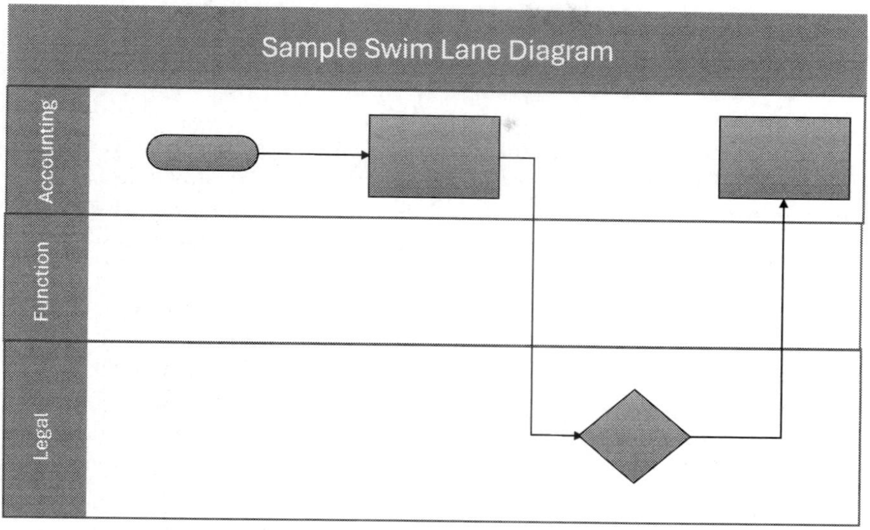

FIGURE 6.3 Swim lane map.

Waste Identification: Helps identify and eliminate waste, reducing costs and improving efficiency.

Improved Communication: Enhances communication and collaboration among team members by providing a common understanding of the process.

Focus on Value: Ensures that efforts are focused on activities that add value to the customer.

Continuous Improvement: Supports ongoing efforts to refine and improve processes over time.

VSM is a powerful tool for organizations aiming to optimize their operations, improve quality, and deliver greater value to customers (Latham & Khosravi, 2022).

LIMITATIONS OF VALUE STREAM MAPPING

VSM also has certain limitations (Teixeira & de Oliveira, 2020):

Data Dependency: The effectiveness of VSM relies heavily on the accuracy and completeness of the data collected by the VSM team. Inaccurate data can lead to misleading maps and poor decision-making.

Static Snapshot: VSM provides a static snapshot of the process at a specific point in time. It does not reflect the dynamic nature of the process that is characterized by ongoing changes and variations.

Complexity Handling: VSM can get quite complicated and challenging when mapping very complex processes involving numerous variables and interactions.

Resource Intensive: The process of creating a detailed VSM can be time-consuming and requires significant training, effort, and the appropriate resources.

Neglects Human and Other Important Factors: VSM primarily focuses on product, material, and information flow, sometimes neglecting other critical aspects like human factors, equipment efficiency, and quality issues.

Implementation Gap: The VSM identifies theoretical improvements; however, practically implementing and sustaining these improvements can be challenging.

Limited to Linear Processes: VSM works best for linear processes and may not be as effective for highly iterative or cyclical processes.

Visioning The Future State: The teams that do not fully comprehend the transformation power of Lean tools cannot envision the future state and struggle to accurately depict it in the VSM. It takes a Lean expert mentor to challenge the team to think beyond the current traditional state of processes and leap into the future Lean state.

Future State Timeline: The timeline for future state maps is not fixed. Usually, it is three to five years. However, for some organizations, six months to one year can be a preferable timeline for the future state. This makes it challenging for teams who need to estimate the extent of transformation within the selected timeline.

CULTURE

Developing a VSM requires significant humanistic intervention from all levels of an organization. As a result of this, it is critical that the two humanistic pillars of Lean – respect and humility are used in the design and development of the current and future state maps.

One of the misconceptions of VSM is that creating a VSM is easy and once created, the future state will be achieved as planned. It is important to understand the success of developing a VSM and executing the steps to get to the expected future state is dependent on the organizational culture (Baitmangalkar, 2021).

Organizations undertaking the Lean transformation journey need to be aware that when drawing a VSM, accounting for culture is crucial for accurately reflecting and improving the processes. Some techniques to incorporate cultural considerations within the VSM process are:

Use VOC and data from the SWOT analysis: Conduct interviews with key personnel to understand the driving and resisting forces within the organization for implementing lean. SWOT analysis feedback from key personnel across the organization can be valuable input for the VSM.

Understand the Norms of the Culture: Identify and respect cultural norms and values that influence behavior and decision-making. Be aware of communication styles, attitudes toward hierarchy, and decision-making processes.

Incorporate Cultural Metrics: Identify and include metrics that reflect cultural dimensions, such as collaboration quality, communication effectiveness, and employee satisfaction. Track these metrics alongside traditional process metrics.

Identify Cultural "Red Flags": It is critical to identify any "Red flags" or warnings within the cultural norms that can impede or prevent the successful implementation of Lean principles. These need to be brought to the attention of the executives and need to be addressed prior to initiating any VSM exercise.

Engage Stakeholders: Involve a diverse group of stakeholders from various cultural backgrounds. Ensure open communication and encourage input from all levels.

Observe Work Practices: Conduct thorough observations to understand how cultural factors affect workflow and interactions. Look for informal practices and rituals that are part of the work culture.

Adapt VSM Symbols and Terminology: Use symbols and terms that are culturally relevant and easily understood by the team. Avoid jargon or concepts that might be unfamiliar or confusing.

Facilitate Inclusive Discussions: Create a safe space for team members to share their perspectives and insights. Use culturally sensitive facilitation techniques to ensure everyone is heard.

Continuous Feedback and Improvement: Establish a feedback loop to continually gather input on how cultural factors are affecting the process. Make adjustments to the VSM as needed to better align with cultural realities.

Training and Awareness: Provide training on cultural awareness and its impact on processes and performance. Foster a culture of continuous learning and adaptation.

By integrating the aforementioned techniques, organizations can create a VSM that not only highlights process inefficiencies but also respects and leverages cultural strengths, leading to more sustainable and effective improvements.

INTEGRATION OF VALUE STREAM MAPPING AND AI

Integrating VSM with AI can address several of the limitations identified above and significantly enhance the process optimization and efficiency of creating the VSM. Here's how this integration can be approached:

Data Collection and Analysis

Automated Data Gathering: AI can automate the collection of real-time data from various sources, ensuring accuracy and reducing the time and effort required for manual data entry. This can be done via sensors, special motion analysis cameras, and other means.

Predictive Analytics: AI algorithms can analyze historical "big" data to predict future trends, identify potential bottlenecks, and suggest proactive measures.

Enhanced Visualization

Dynamic Mapping: AI can enable dynamic, real-time updates to VSMs, providing a live view of the process and its performance.

Pattern Recognition: AI can detect patterns and anomalies in the data that may not be apparent through manual analysis, offering deeper insights into process inefficiencies.

Process Optimization

Advanced Simulations: AI can simulate various scenarios to test the impact of different changes without disrupting the actual process. This helps in identifying the best course of action for process improvement.

Optimal Pathways: AI can analyze multiple process pathways and recommend the most efficient routes for material and information flow, optimizing resource allocation and minimizing waste.

Continuous Improvement

Feedback Loops: AI can create continuous feedback loops by monitoring process performance and automatically updating the VSM with new data, ensuring ongoing improvements.

Adaptive Learning: AI systems can learn from past data and continuously improve their recommendations, adapting to changing conditions and evolving process requirements.

Human-AI Collaboration

Augmented Decision-Making: AI can provide actionable insights and recommendations, empowering human decision-makers with data-driven support while allowing them to focus on strategic and creative aspects of process improvement.

Skill Enhancement AI can assist in training and upskilling employees by providing them with real-time insights and guidance based on the VSM data.

Integrating VSM with AI can create a more robust, responsive, and efficient process improvement framework, driving significant gains in productivity and competitiveness.

INTEGRATING VSM WITHIN THE LEAN-AI MODEL

VSM is best applied within an organization at strategic, operational, and tactical levels as shown in Figure 6.4. At the start of the Lean journey, a strategic VSM-AI can be developed using the executive team that can help get a high-level understanding of the current state of the process, material, and information flow within the organization. This helps identify the improvement opportunities and kaizen and kaikaku events required to get to the future "ideal" state. Once a clear picture is available of the overall journey, specific areas within the strategic VSM-AI can be highlighted for targeted improvements using kaizen events. During the kaizen events, middle management can guide the creation of a tactical VSM-AI of the highlighted target area and the team can address the identified inefficiencies. An operational VSM-AI is used by the team members trained on VSM on an ad-hoc basis to understand and solve any problems or failures that may occur in the process. This helps identify the

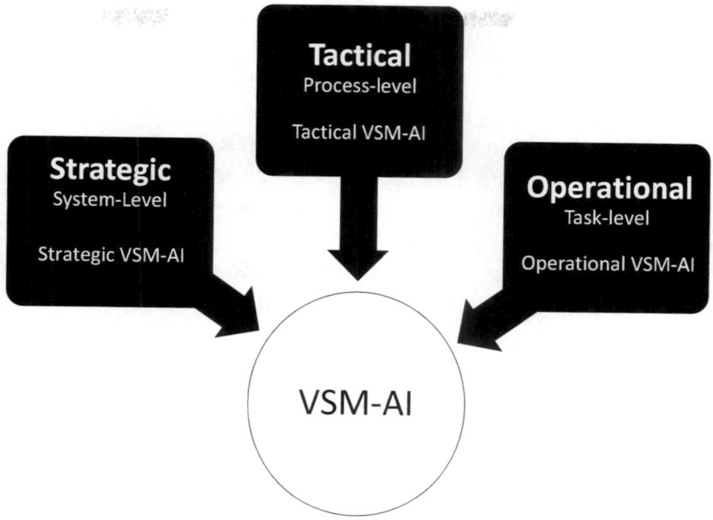

FIGURE 6.4 VSM-AI model.

root causes and helps in developing a new standard process that prevents any recurrence of the failures.

The integrated VSM-AI model provides specific tactical, operational, and strategic actions to ensure that VSM provides optimal results for the organization as shown in Figure 6.4.

VSM-AI DATA PRIVACY AND ETHICS CONSIDERATIONS

Implementing VSM with AI in organizations can drive significant efficiency gains but also brings data privacy and ethical challenges that must be navigated carefully. As VSM shifts from traditional manual mapping to digital and AI-enhanced analysis, the following considerations are key to balancing operational efficiency with ethical responsibility:

Data Collection, Transparency, and Consent: Clearly communicate the types of data collected during VSM processes, particularly when this data includes employee interactions, task performance, and resource utilization. Obtain consent from all employees and stakeholders involved, specifying how their data will contribute to the VSM and broader process improvements, which enhances transparency and trust.

Data Minimization and Purpose Limitation: Collect only the data necessary for achieving the AI-driven VSM objectives, such as streamlining processes, identifying bottlenecks, and enhancing workflow efficiencies. Ensure data usage is strictly limited to the intended purpose to avoid function creep, where collected data might be repurposed for other applications without proper disclosure or consent.

Security and Data Protection: Since VSM data can include sensitive operational details and employee performance metrics, implement strong data protection measures, including encryption, access control, and frequent security audits. Protecting

VSM data from breaches or unauthorized access is essential, especially when it could reveal competitive or private information.

Bias and Fairness in AI Algorithms: AI algorithms used in VSM should be regularly audited to ensure they remain fair and unbiased, especially when impacting workflow decisions that could affect employees differently based on role, location, or department. Monitor for and mitigate any bias that could lead to unfair treatment, such as favoring certain employee groups for efficiency gains or unfairly penalizing others for bottlenecks.

Employee Privacy and Anonymization/Coding: Consider anonymizing, aggregating, or coding data related to individual employees whenever possible. While VSM aims to improve overall workflow, it's important to avoid excessive monitoring of individuals. Anonymization protects privacy while still allowing AI to analyze process flows without identifying specific employees' productivity or performance.

Transparency and Explainability in AI Recommendations: Ensure AI outputs are understandable to managers and employees, especially when VSM insights impact daily workflows, process roles, or performance expectations. Make AI-based recommendations explainable and accessible, allowing employees to understand the reasoning behind workflow changes or efficiency suggestions derived from AI.

Human Oversight and Accountability: Adopt a human-in-the-loop model where AI insights generated from VSM are reviewed by process owners and stakeholders before implementing major changes. Assign accountability for AI-driven decisions to a dedicated governance team, which ensures that the ethical and privacy implications of AI are continuously reviewed and aligned with company values.

Engaging Employees in the AI-Driven VSM Process: Involve employees in the process of VSM enhancement using AI by providing training, soliciting feedback, and addressing concerns proactively. Employee engagement promotes ethical, inclusive implementation and builds a culture of trust, which is critical when transitioning from manual to AI-enhanced VSM.

Potential Impact on Employment and Job Roles: Be transparent about how AI-driven VSM could impact roles and responsibilities, especially if it reveals redundancies or process adjustments. Maintain an ethical commitment to reskilling and upskilling employees affected by AI-driven changes, balancing efficiency with a commitment to workforce development.

Establishing Governance and Ethical Review Boards: Create a dedicated team to oversee the ethical use of AI in VSM, implementing policies and procedures to guide data privacy and usage consistently. Ethical review boards can assess the implications of new AI tools in VSM, ensuring all enhancements align with company ethics, data protection standards, and regulatory requirements.

By proactively addressing these considerations, organizations can responsibly leverage AI in VSM while safeguarding employee privacy and building an ethical foundation for sustainable operational improvements. This approach not only maximizes the potential of AI in VSM but also reinforces trust and alignment with organizational values.

VSM-AI INTEGRATION INDUSTRY EXAMPLES

Integrating VSM with AI technology can significantly enhance process efficiency and decision-making across various industries. Next are some examples of applications of VSM-AI.

MANUFACTURING

Predictive Maintenance:

Example: A car manufacturing plant uses AI-driven predictive maintenance to analyze machine data in real time. By integrating this with VSM, they can identify potential bottlenecks due to machine downtime and proactively schedule maintenance to minimize disruptions.

Quality Control:

Example: A semiconductor manufacturing company uses AI for real-time defect detection and quality control. Combining this with VSM allows them to map out the defect detection process and identify stages where defects are most likely to occur, leading to targeted improvements.

HEALTHCARE

Patient Flow Optimization:

Example: A hospital integrates AI with VSM to analyze patient flow data and predict peak times in emergency rooms. This helps in optimizing staffing levels and resource allocation, reducing wait times, and improving patient care.

Supply Chain Management:

Example: A pharmaceutical company uses AI to forecast demand for medications and integrate this data with VSM. This helps in identifying inefficiencies in the supply chain and ensures timely delivery of critical medications.

RETAIL

Inventory Management:

Example: A retail chain integrates AI with VSM to optimize inventory levels. AI algorithms predict product demand, which is then mapped onto the VSM to streamline restocking processes and reduce excess inventory.

Customer Service:

Example: An e-commerce company uses AI chatbots to handle customer inquiries and integrates this with VSM to streamline customer service processes. The VSM helps identify stages where AI can take over routine tasks, allowing human agents to focus on more complex issues.

LOGISTICS AND TRANSPORTATION

Route Optimization:

Example: A logistics company integrates AI with VSM to optimize delivery routes. AI analyzes traffic patterns and delivery schedules, and VSM maps out the most efficient routes, reducing delivery times and fuel costs.

Warehouse Management:

Example: A distribution center uses AI-powered robots for order picking and integrates this with VSM. The VSM helps in identifying inefficiencies in the picking process and AI provides real-time data to improve order accuracy and speed.

FINANCIAL SERVICES

Fraud Detection:

Example: A bank integrates AI with VSM to streamline the fraud detection process. AI algorithms analyze transaction data to detect anomalies, and VSM maps out the investigation process to improve response times and reduce false positives.

Loan Processing:

Example: A financial institution uses AI to assess loan applications and integrates this with VSM. The VSM helps identify stages where AI can automate tasks, such as initial credit checks, and speeding up the loan approval process.

By integrating AI with VSM, these industries can leverage real-time data and predictive analytics to streamline processes, reduce inefficiencies, and enhance overall performance.

VSM-AI ADVANTAGES

Integrating VSM and AI technology offers several advantages that can lead to significant improvements in efficiency, decision-making, and overall process performance. Here are some key benefits:

ENHANCED DATA ANALYSIS

Real-Time Insights: AI provides real-time data analysis, enabling dynamic updates to the VSM, which helps in identifying and addressing issues as they occur.

Predictive Analytics: AI algorithms can predict future trends and potential bottlenecks, allowing for proactive adjustments in the value stream.

IMPROVED EFFICIENCY

Automated Processes: AI can automate repetitive and time-consuming tasks, freeing up human resources for more strategic activities and improving overall efficiency (Silva & de Sousa, 2019).

Optimal Resource Allocation: AI helps in optimal resource allocation by analyzing patterns and suggesting the best ways to deploy resources across the value stream.

BETTER DECISION-MAKING

Data-Driven Decisions: Integrating AI with VSM provides a robust foundation for data-driven decision-making, reducing reliance on intuition and improving accuracy.

Enhanced Visualization: AI can enhance the visualization of complex data within the VSM, making it easier to understand and act upon.

INCREASED AGILITY

Rapid Adaptation: AI enables faster adaptation to changes in the market or operational environment by quickly identifying and implementing necessary adjustments in the value stream.

Continuous Improvement: AI facilitates continuous improvement by constantly analyzing data and providing insights for incremental changes, ensuring the value stream remains efficient and effective.

COST REDUCTION

Reduced Operational Costs: By identifying inefficiencies and automating tasks, AI helps in reducing operational costs and improving profitability.

Minimized Downtime: Predictive maintenance and real-time monitoring through AI can minimize downtime, ensuring a smoother and more consistent value stream.

ENHANCED QUALITY AND COMPLIANCE

Improved Quality Control: AI enhances quality control processes by providing precise and consistent monitoring, reducing defects, and ensuring higher product quality.

Regulatory Compliance: AI helps in maintaining regulatory compliance by continuously monitoring processes and ensuring they meet the required standards.

INCREASED CUSTOMER SATISFACTION

Better Customer Experience: By optimizing processes and reducing lead times, AI integration can lead to improved customer satisfaction and loyalty.

Personalized Services: AI can analyze customer data to provide personalized services and recommendations, enhancing the overall customer experience.

COMPETITIVE ADVANTAGE

Innovation and Differentiation: Integrating AI with VSM can drive innovation, offering a competitive edge by implementing cutting-edge technologies in process management.

Scalability: AI enables scalable solutions, allowing businesses to efficiently expand their operations without a proportional increase in costs or complexity.

Automated Processes: AI can automate repetitive and time-consuming tasks, freeing up human resources for more strategic activities and improving overall efficiency.

Optimal Resource Allocation: AI helps in optimal resource allocation by analyzing patterns and suggesting the best ways to deploy resources across the value stream.

BETTER DECISION-MAKING

Data-Driven Decisions: Integrating AI with VSM provides a robust foundation for data-driven decision-making, reducing reliance on intuition and improving accuracy.

Enhanced Visualization: AI can enhance the visualization of complex data within the VSM, making it easier to understand and act upon.

INCREASED AGILITY

Rapid Adaptation: AI enables faster adaptation to changes in the market or operational environment by quickly identifying and implementing necessary adjustments in the value stream.

Continuous Improvement: AI facilitates continuous improvement by constantly analyzing data and providing insights for incremental changes, ensuring the value stream remains efficient and effective.

COST REDUCTION

Reduced Operational Costs: By identifying inefficiencies and automating tasks, AI helps in reducing operational costs and improving profitability.

Minimized Downtime: Predictive maintenance and real-time monitoring through AI can minimize downtime, ensuring a smoother and more consistent value stream.

ENHANCED QUALITY AND COMPLIANCE

Improved Quality Control: AI enhances quality control processes by providing precise and consistent monitoring, reducing defects, and ensuring higher product quality.

Regulatory Compliance: AI helps in maintaining regulatory compliance by continuously monitoring processes and ensuring they meet the required standards.

INCREASED CUSTOMER SATISFACTION

Better Customer Experience: By optimizing processes and reducing lead times, AI integration can lead to improved customer satisfaction and loyalty.

Personalized Services: AI can analyze customer data to provide personalized services and recommendations, enhancing the overall customer experience.

COMPETITIVE ADVANTAGE

Innovation and Differentiation: Integrating AI with VSM can drive innovation, offering a competitive edge by implementing cutting-edge technologies in process management.

Scalability: AI enables scalable solutions, allowing businesses to efficiently expand their operations without a proportional increase in costs or complexity.

By leveraging these advantages, organizations can achieve greater efficiency, effectiveness, and adaptability, ultimately driving better performance and higher satisfaction for both customers and stakeholders.

VALUE STREAM ANALYSIS USING CHATGPT

An evaluation of a sample current state VSM of gear manufacturing operations, as shown in Figure 6.5, demonstrates the power of ChatGPT to qualitatively and quantitatively evaluate the map (ChatGPT, n.d.):

The overview of the VSM is given here (Cudney, 2009):

The raw material is shipped in by truck

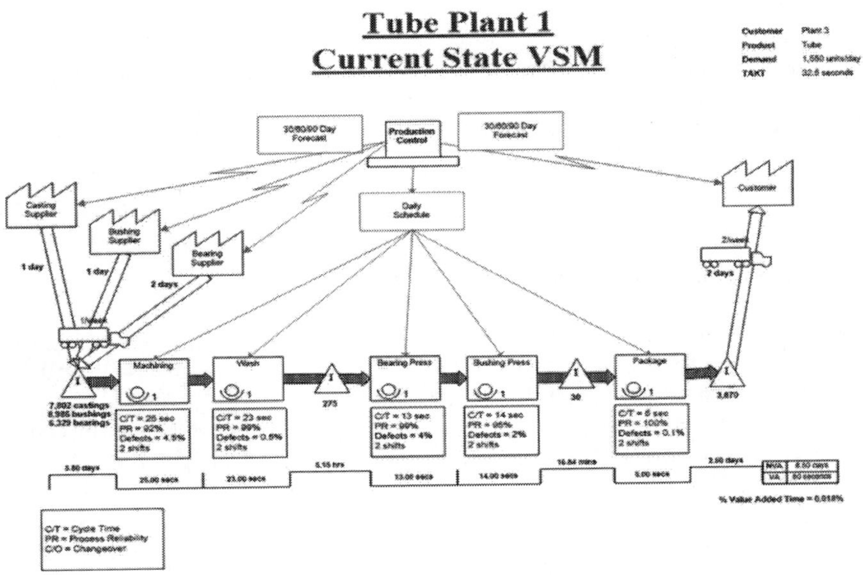

FIGURE 6.5 Sample current state value stream map (Cudney, 2009).

Shipments are received once a week, usually on Fridays to ensure the raw material is received prior to the start of the next production week
Shipments from the casting supplier and bearing supplier take 1 day
Shipments from the bushing supplier take 2 days
Lead time is 8.8 days
No changeover is required
Process reliability is 84%
Yield is 80%
The first pass yield is 63%

Observed inventory:

7,802 tubes (in receiving)
8,986 bushings (in receiving)
6,329 bearings (in receiving)
275 tubes machined and washed (ready for bearing press)
30 final products waiting for packaging
3,870 tubes packaged and waiting for shipment to the customer

Tubes are shipped by truck and the deliveries occur twice per week and take 2 days.

For the gear housing assembly, the daily customer demand is 775 units. All three plants run a two-shift operation.

After uploading the sample VSM, the following are the prompts to and responses from ChatGPT.

ChatGPT Prompt: " Evaluate This VSM"
ChatGPT Response:

Looking at the Value Stream Map (VSM) for Tube Plant 1, here are several observations and improvement suggestions:

High Inventory Levels

Observation: There are significant inventories (7,802 castings, 8,896 bushings, 6,329 bearings) waiting in the pipeline. Additionally, there are 275 units waiting after the washing stage, 30 units after the bushing press, and 3,870 units before packaging.

Improvement: Implement a Just-in-Time (JIT) system or Kanban to reduce the amount of Work-in-Progress (WIP) inventory. This can help to streamline production and reduce waste. A more frequent delivery schedule from suppliers could also reduce incoming inventory and avoid overproduction.

Long Lead Times

Observation: The lead time of 8.5 days is significantly higher than the value-added time of just 80 seconds, indicating that most time is spent in non-value-adding activities (NVA).

Improvement: Focus on reducing waiting times between processes and improving flow. Value Stream Segmentation or Process Synchronization may help align the processes better and eliminate excessive delays.

Takt Time vs. Cycle Time

Observation: The TAKT time is 32.5 seconds, but many cycle times are shorter (e.g., 5 seconds at packaging, 13–25 seconds in other stages). However, there's no immediate bottleneck indicated.

Improvement: Ensure that the cycle times are optimized and aligned with the TAKT time. If some stations are consistently ahead of schedule, they might be overproducing, contributing to higher WIP. Introducing a Level Production Plan (Heijunka) could balance production and prevent this overproduction.

Process Reliability and Defects

Observation: While most process reliability (PR) is quite high (above 90%), the machining process has the lowest reliability (92%) and the highest defect rate (4.5%). Bearing Press has a 4% defect rate as well.

Improvement: Investigate the root cause of defects in the machining and bearing press. Implement Total Productive Maintenance (TPM) to increase machine reliability. Introducing error-proofing (Poka-Yoke) could further reduce defect rates in these areas.

Opportunities to Reduce Non-Value-Added Time

Observation: There is a large amount of non-value-added (NVA) time in the system (8.5 days). The packaging step has only 5 seconds of value-added time but large inventory waiting beforehand.

Improvement: Map out the causes of NVA activities, focusing on logistics, material handling, and scheduling, and optimize these areas. Implementing continuous flow in areas where feasible or cellular manufacturing could also help minimize waiting and transportation.

Supplier Delivery Frequency

Observation: Suppliers are delivering once a week, which leads to high inventory levels and increases the risk of overproduction.

Improvement: Work with suppliers to increase delivery frequency, possibly to daily or every other day deliveries, using a pull system to match the exact demand of production.

Customer Delivery Cycle

Observation: Customer shipments occur twice a week, with a two-day lead time for shipping.

Improvement: Consider aligning the production pace more closely with customer demand, potentially reducing the batch size shipped and shifting towards smaller, more frequent deliveries if feasible. This would reduce the inventory buildup before packaging.

ChatGPT Prompt: "Suggest Improvements for This VSM"

ChatGPT Response:

Key Focus Areas for Improvement:

Reduce Inventory and Lead Time: Minimize waiting times and excessive WIP through better production scheduling and more frequent supplier deliveries.

Increase Process Reliability: Target improvement in processes like machining and bearing press with higher defect rates.

Align Production to Demand: Use Heijunka to balance production with Takt time and avoid overproduction.

Reduce Non-Value-Added Time: Address waiting and material movement inefficiencies to increase the value-added ratio.

By implementing these changes, the plant could reduce waste, improve flow, and enhance overall efficiency.

ChatGPT Prompt: "For the Gear Housing Assembly, the Daily Customer Demand Is 775 Units and a Two-Shift Operation. There Is a 45-Minute Break in Each Shift. Calculate the Takt Time."

ChatGPT Response:

If each shift has a 45-minute break, we need to adjust the available production time accordingly.

Step 1: Adjust available production time

- Total shift duration = 8 hours = 480 minutes
- Break per shift = 45 minutes
- Net production time per shift = 480 = 45 = 435 minutes Now, let's calculate the available production time for both shifts:

Available time (seconds/day) = 2 shifts x 435 minutes/shift x 60 seconds/minute = 52,200 seconds/day

Step 2: Recalculate TAKT time

Using the adjusted available production time:

$$\text{TAKT time} = \frac{52,200 \text{ seconds / day}}{775 \text{ units / days}} = 67.35 \text{ seconds / unit}$$

Final Answer:

With a 45-minute break per shift, the TAKT time is 67.35 seconds per unit.

Chat-GPT Prompt: "What Is the Value-Added Percentage?"

ChatGPT Response:

To calculate the Value-Added Percentage, we need the Value-Added Time (VA) and the Total Lead Time (including Non-Value-Added Time, NVA).

Formula:

$$\text{Value - Added Percentage} = \left(\frac{\text{Value - Added Time (VA)}}{\text{Total Lead Time (VA + NVA)}} \right) \times 100$$

From the provided VSM:

- *Value-Added Time (VA) = 80 seconds*
- Non-Value-Added Time (NVA) = 8.50 days = $8.50 \times 24 \times 60 \times 60$ = 734,400 seconds

Total Lead Time:

Total Lead Time = 80 seconds (VA) + 734,400 seconds (NVA) = 734,480 seconds

Calculate Value-Added Percentage

$$\text{Value-Added Percentage} = \left(\frac{80}{734,480} \right) \times 100 = 0.109\%$$

Final Answer:

The value-added percentage is approximately 0.011%.

This exercise can be repeated with other AI tools like Microsoft Copilot and others. This simple exercise provides an insight into the power of AI tools in evaluating a VSM.

CONCLUSION

A VSM is the graphical documentation of all the actions (value-added, necessary non-value-added, and non-value-added) that occur to a product or service, beginning with its inception (at order) through shipment and delivery to the customer. It is a visual tool used in Lean methodology to map out, analyze, and design the flow of processes, materials, and information required to bring a product or service to a customer.

However, developing a VSM is time-consuming, data-dependent, resource-intensive, static visual at a moment in time, and neglects human factors. Integrating AI with VSM results in enhanced data analysis, real-time insights, predictive analytics, improved efficiency, and better data-driven decision-making leading to increased agility and efficiency of VSM creation and execution.

EXERCISES

1. What are the current system and process limitations of your VSM approach, and how can you apply AI to overcome these limitations?

VSM Step	System and Process Limitations	AI Approach to Overcome System and Process Limitations
Step 1: Determining the customer requirements		
Step 2: Identifying the process steps (from supplier to customer)		
Step 3: Identifying the material flow (from supplier to customer)		
Step 4: Identifying the information flow (from supplier to customer)		
Step 5: Identifying the cycle time, lead time, and changeover times for all the process steps		
Step 6: Identifying the value-added and non-value-added times for all the process steps		

2. What are the current cultural limitations of your VSM approach, and how can you apply AI to overcome these limitations?

VSM Step	Cultural Limitations	AI Approach to Overcome Cultural Limitations
Step 1: Determining the customer requirements		
Step 2: Identifying the process steps (from supplier to customer)		
Step 3: Identifying the material flow (from supplier to customer)		
Step 4: Identifying the information flow (from supplier to customer)		
Step 5: Identifying the cycle time, lead time, and changeover times for all the process steps		
Step 6: Identifying the value-added and non-value-added times for all the process steps		

3. What are some short- and long-term benefits to your department and organization for integrating AI into VSM?

#	Short-Term Benefits	Long-Term Benefits
Example	*Increased data accuracy*	*Improved quality of VSM*
1		
2		
3		
4		
5		
6		

4. Reflect on the ethical implications of using AI in VSM within your organization. Consider aspects such as employee privacy, job displacement, and data security. Develop strategies to mitigate the negative aspects of the workforce.

#	Ethical Considerations	Strategies to Mitigate
Example	*Job displacement fears*	• *Provide upskilling and reskilling programs* • *Clear communication about AI's role*
1		
2		
3		
4		
5		
6		

5. What are the technology considerations for integrating AI into your VSM framework within your organization? Develop strategies for the necessary technology integration.

#	Technology Necessary for AI	Strategy for Integration
Example	*Real-time data analytics*	• *Develop robust data infrastructure* • *Train staff on data management*
1		
2		
3		
4		
5		
6		

REFERENCES

Baitmangalkar, S. (2021). The delusion of value stream mapping. *LinkedIn*. https://www.linkedin.com/pulse/delusion-value-stream-map-sanjeev-baitmangalkar/

ChatGPT. (n.d.). Value stream mapping: Process flow tools. *ChatGPT*. https://chatgpt.com/g/g-tCIPTBaaG-value-stream-mapping-process-flow-tools

Cudney, E. (2009). *Using hoshin kanri to improve the value stream*. New York, NY: CRC Press.

Duflou, J. R., & De Meulemeester, J. (2021). A literature review on value stream mapping: Current developments and future perspectives. *Production Planning & Control*, *32*(1), 43–55. https://doi.org/10.1080/09537287.2020.1806547

Hines, P., & Rich, N. (1997). The seven value stream mapping tools. *International Journal of Production Economics*, *45*(1–3), 39–58. https://doi.org/10.1016/S0925-5273(96)00023-5

Latham, J., & Khosravi, H. (2022). Using swim lane mapping to enhance operational efficiency in manufacturing. *Production Planning & Control*, *33*(10), 847–856. https://doi.org/10.1080/09537287.2022.207200

Melton, T. (2022). Value stream mapping: A method for integrating quality and efficiency. *Quality Management Journal*, *29*(1), 55–67. https://doi.org/10.1080/10686967.2022.2011253

Rother, M., & Shook, J. (2003). *Learning to see: Value stream mapping to add value and eliminate Muda*. Lean Enterprise Institute.

Rumbaugh, J., & Blaha, M. (2019). *A practical guide to process modeling using swim lane diagrams*. Software Engineering Institute. https://resources.sei.cmu.edu/asset_files/TechnicalReport/2019_005_001_64270.pdf

Sharma, S., & Sharma, R. (2020). Value stream mapping: A tool for improvement in manufacturing. *International Journal of Quality & Reliability Management*, *37*(5), 657–671. https://doi.org/10.1108/IJQRM-03-2019-0074

Silva, E. J. D., & de Sousa, R. A. (2019). The impact of value stream mapping on operational performance: A study in Brazilian companies. *International Journal of Lean Six Sigma*, *10*(2), 562–581. https://doi.org/10.1108/IJLSS-06-2018-0066

Teixeira, F. S., & de Oliveira, A. F. (2020). Value stream mapping and improvement: A case study in a manufacturing company. *Procedia Manufacturing*, *51*, 377–384. https://doi.org/10.1016/j.promfg.2020.10.053

7 Poka-Yoke-AI

POKA-YOKE OVERVIEW

Poka-yoke or mistake proofing was pioneered by Shigeo Shingo. Shingo was an industrial engineer for the TPS in the 1960s. It was originally known as "baka-yoke" or "idiot-proofing." The intent was that any mistake made by an individual should not make them look or feel like "an idiot." Although Shigeo's intent was pure and aimed at preserving human dignity. Due to some operators being offended by the term "baka" (idiot), Shigeo changed it to "poka-yoke," or "error-proofing."

Shingo invented poka-yoke when during an assembly process of a switch button at a Toyota factory, the operators invariably forgot to install the spring below the switch. Shingo reengineered the process to be a two-step assembly; the first step required the operator to place the springs in a placeholder, and the second step required them to insert the springs into the switch. If a spring was still visible in the placeholder, it would be obvious to the operator that they had forgotten to place it in the switch. This would allow them an opportunity to correct their mistake immediately, preventing the assembly of a defective switch. Thus, poka-yoke, born as a trouble-shooting technique, became an integral and invaluable tool in the Lean toolbox. Without poka-yoke, it would be difficult to maintain other Lean tools like JIT, kanban, and others (Shingo, 1986).

One of the key aspects of poka-yoke is that the solutions are generally creative, simple, and inexpensive. These solutions are not limited to manufacturing alone but can be applied to daily lives. An example is a facial tissue box. How is the user supposed to know that they are reaching the end of the stack of white tissues that are not visible from outside the enclosed box? The solution is simple. The end of the tissue stack is of a different color – as soon as the new color tissue shows up, it signals nearing the end of the tissue stack. This makes poka-yoke a simple and inexpensive solution that can help humans prevent mistakes in our daily lives.

A poka-yoke tool is usually a tactical tool when applied to products and processes, however, when applied to organizational and system-level issues, it can be a strategic advantage. For example, in the case of vehicles, when poka-yoke is applied to the dashboard controls to alert the driver to a low oil level or overheating., it is a tactical application to safeguard the driver from making mistakes and preventing damage to certain components of the vehicle. When poka-yoke is integrated with a facial recognition camera system to prevent accidents with children and pets while backing up vehicles, it becomes a strategic application to prevent accidents and fatalities.

DOI: 10.1201/9781003532972-7

CATEGORIES OF POKA-YOKE

Human oversight causes errors, and human errors tend to result in defects. Mistakes will not turn into defects if errors are discovered & eliminated before they occur. The key to effective poka-yoke is to prevent errors from becoming defects. Poka-yoke techniques can be classified into five categories that help accomplish this goal:

Elimination: The elimination type of poka-yoke eliminates any possibility of a mistake by designing out any possibility for an error. For example, the auto-braking feature in a vehicle automatically detects the distance from vehicles ahead and based on calculations of the relative speeds of the vehicles, applies the appropriate braking pressure to bring the vehicle to a safe stop to eliminate the possibility of a head-on collision.

Prevention: The prevention type of poka-yoke helps humans avoid making the mistake. For example, due to the design of the various electric sockets based on international standards, a 110 V US-based electric plug cannot be inserted into a 220 V non-US socket, which could otherwise result in serious damage to the attached device.

Detection: The detection type of poka-yoke detects and alerts humans when mistakes are made. This prevents mistakes from becoming defects. For example, if an iron is left on by mistake, it can have an automatic shut-off function wherein the iron will turn off after 15 minutes if left in the upright position or after 30 seconds if left on its soleplate or its side.

Correction: The correction type of poka-yoke detects and automatically corrects the mistakes without any human intervention. For example, the autosave feature in computer applications or the turn indicator auto-shut-off feature in vehicles after the turn is completed.

Loss Control: The loss control type of poka-yoke is used to minimize the damage from defects when the potential for defects to occur is high, and the ability to detect the defects is low. For example, ground fault circuit interrupters (GFCIs) in eclectic circuits are designed to break the circuit as soon as it detects that the electrical load on the circuit exceeds the designed load capacity, thus preventing electric shocks to humans and damage to the equipment.

DECISION DIAGRAM FOR AN EFFECTIVE POKA-YOKE

Figure 7.1 shows a flowchart to make decisions regarding using the appropriate poka-yoke for the right problem.

ADVANTAGES OF POKA-YOKE

The objective of error-proofing is zero defect. Since human mistakes are inevitable, poka-yoke is a proactive and inexpensive system to avoid mistakes that become defects, thus benefiting both internal and external customers. The primary advantage of implementing the poka-yoke tool in Lean manufacturing is that it eliminates, reduces, or prevents mistakes thus eliminating or reducing defects. In addition,

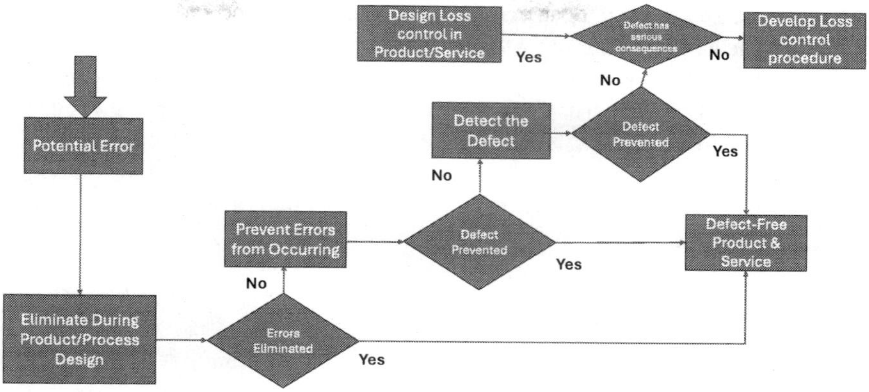

FIGURE 7.1 Poka-yoke decision diagram.

poka-yoke tools are generally simple, inexpensive, and can be easily incorporated within existing processes.

Another benefit of poka-yoke is greater safety for employees and consumers resulting in lower worker compensation, accident-related expenses, and greater confidence among management, employees, and customers. Additional benefits of poka-yoke are increased flexibility, lower manufacturing/service costs, and greater commitment to continuous improvement at all levels of the organization. The elimination, reduction, or prevention of errors and subsequent customer confidence (both internal and external) result in lower prices, superior products, better customer satisfaction, enhanced loyalty, and higher profits (Lazarevic et al., 2019).

POKA-YOKE: ENHANCING THE BENEFITS OF LEAN TOOLS/TECHNIQUES

Elimination of Waste (Muda): Elimination of waste is the fundamental requirement of lean. All defects, rework, and corrective actions are forms of waste that violate the basic Lean principle of doing things right the first time. Poka-yoke provides some guardrails and eliminates or reduces the violations.

Single-Piece Flow: Some of the fundamental benefits of Lean come from the concept of a single-piece flow. The success of two pillars of lean, namely kanban and JIT, depend on the implementation of the one-piece flow. The presence of errors and defects disrupts the single-piece flow resulting in a breakdown of kanban and JIT and subsequently an inability to get the full benefits of Lean technology. Therefore, without poka-yoke, the efficacy of Lean is significantly lower.

Minimal Inventory: One of the precepts of Lean is minimal inventory as inventory is considered to be one of the eight wastes of Lean. When the products or processes have defects, it can result in stoppages or shut downs of lines. Poka-yoke helps eliminate or reduce the defects that can prevent shutdowns or reduce stoppages.

Equipment Readiness: Lean requires that the equipment and processes are available and ready for use when needed, without which the OEE is significantly reduced. When the equipment, products, or processes have defects, it can result in stoppages or shut downs of lines. Poka-yoke helps eliminate or reduce the defects that can improve availability, prevent shutdowns, or reduce stoppages. Mistake-proofing processes, tools, gauges, fixtures, documents, methodologies, and systems ensure optimal availability resulting in better OEE.

LIMITATIONS OF POKA-YOKE

Like any tool, poka-yoke also has some limitations that prevent it from being effective, efficient, and widely used across several applications. Limitations range from the data collection process to the culture of the organization.

Data collection: To design an accurate and reliable poka-yoke system, it is important to collect accurate and reliable defect-related data from internal and external customers. The better the data and analysis thereof, the more effective will be the poka-yoke systems in preventing human errors.

The complexity of Design and Implementation: Designing and Implementing poka-yoke systems can be quite complicated when the process is complex, has several variables, and is prone to many human errors. Designing and implementing a foolproof poka-yoke system requires detailed planning and skilled personnel.

Cost of Design and Implementation: Some poka-yoke systems may need to be complex and sophisticated to prevent human errors in critical products and processes. The design, implementation, and training required for such systems can be cost-prohibitive for some small- and medium-sized companies as they may lack the technical expertise, resources, time, and financial resources needed for this upgrade and future maintenance of the implemented systems.

Stakeholder Resistance: Implementing poka-yoke systems requires updates to the workflow, and procedures, and may also require personnel to be trained on the new approach. This can result in resistance to change from employees who are used to existing processes and methodologies. There may also be pushback from the decision-makers as they may not be convinced of the payback. Without the required buy-in from all stakeholders, overcoming this resistance to change can be challenging.

Ineffective Solutions: Designing and implementing poka-yoke systems is an art and a science. There is no single approach to this, and each solution needs to be tailored to the requirements.

Over-reliance on Poka-yoke Systems: The personnel risk becoming complacent if they over-rely on mistake-proofing systems. Over-dependence on poka-yoke to catch all errors makes employees less vigilant leading to other types of mistakes.

False Sense of Security: Poka-yoke systems are not foolproof solutions to prevent all errors. The possibility of new, unexpected errors occurring exists, resulting in a false sense of security among the stakeholders. If the poka-yoke system fails for some reason, it can result in serious defects and recalls.

Need for Calibration and Maintenance: All poka-yoke systems require regular calibration and maintenance to remain effective in preventive errors. Lack of attention to calibration, maintenance, and upkeep of these systems can result in deterioration of effectiveness over time to prevent human errors.

Impact on Culture: An unintended result of over-reliance on poka-yoke systems may be a lack of creativity and problem-solving skills among employees. It may also impact teamwork and eliminate the need for teams to improve communication skills.

CULTURE

Since the application of poka-yoke is a human-centric process, during developing a poka-yoke, the focus on the two humanistic pillars of Lean – respect and humility are critical to the planning, execution, and sustainment of the poka-yoke initiatives.

As with any methodology, there are limitations from a culture and workforce perspective that must be considered. These limitations often stem from resistance to change, over-reliance on systems, perception of micromanagement, and a culture that lacks accountability.

First, employees may resist the implementation of poka-yoke systems due to fear of losing autonomy or jobs, particularly if automation or mistake-proofing mechanisms seem to replace human judgment. This can create friction, especially in organizations with strong traditional practices (Lazarević et al., 2023). Secondly, while poka-yoke aims to prevent errors, over-reliance on such systems can diminish critical thinking and problem-solving skills. Employees may become too dependent on the tools rather than addressing root causes of issues, leading to a decline in innovation and proactive problem-solving within the workplace (Shingo, 1986). Thirdly, some employees might feel that poka-yoke introduces unnecessary oversight or controls, especially in environments where trust between leadership and workers is crucial. If not communicated properly, this can lead to lower morale and a sense of micromanagement. Also, by preventing errors, poka-yoke systems can inadvertently shift the focus away from personal accountability, where employees may start to rely on the system to catch mistakes rather than taking ownership of quality control.

While the aforementioned are the negative impact on culture, but on the positive side, successfully integrating poka-yoke can foster a culture of continuous improvement (kaizen), where employees feel empowered to spot and address inefficiencies. This encourages a mindset where mistakes are viewed as learning opportunities, contributing to a collaborative and open culture (Hines & Rich, 1997).

INTEGRATION OF POKA-YOKE AND AI

As seen above, there are significant limitations to poka-yoke systems. Due to these limitations, poka-yoke systems fall short of the intended goal of zero defects.

Integration of poka-yoke and AI is a pioneering effort that can help resolve some of the limitations and help attain the objective of zero defects.

Lean and AI, when used together, can transform poka-yoke systems into more effective, accurate, reliable, and less expensive with automated features.

Integrating poka-yoke (mistake-proofing) with AI can help resolve several cultural issues by enhancing efficiency, reducing fear of change, and fostering a more positive work environment.

For example, AI-powered poka-yoke systems can be designed to assist employees rather than replace them. By positioning AI as a support tool that enhances human capabilities, organizations can reduce the fear of job displacement and resistance to change. AI can help prevent errors without taking over decision-making, allowing workers to focus on more value-added tasks (Lazarević et al., 2023). Poka-yoke and AI could automate repetitive tasks while providing real-time feedback and learning opportunities. This encourages employees to develop new skills and remain engaged in critical thinking, rather than becoming overly reliant on mistake-proofing mechanisms. AI's ability to offer insights into process improvement also fosters continuous learning and innovation (Rother & Shook, 2003). AI-driven poka-yoke systems can analyze vast amounts of data to identify patterns and opportunities for process optimization. This supports a kaizen (continuous improvement) culture by empowering employees with data-driven insights. Workers can proactively contribute to process enhancements, strengthening collaboration and engagement (Hines & Rich, 1997). Another important impact on culture could be that AI can help mitigate concerns about micromanagement by providing objective, real-time monitoring of processes. This can shift the focus from blame to improvement, promoting a culture of trust. Employees see AI as a tool for maintaining quality and safety, rather than as an oversight mechanism. Also, by integrating AI, poka-yoke systems can offer personalized feedback, helping employees take ownership of their tasks. AI-driven mistake-proofing enhances personal accountability without creating an over-reliance on systems, as workers still remain responsible for ensuring quality control.

In summary, AI-integrated poka-yoke systems not only prevent errors but also support a more dynamic and inclusive workplace culture by reducing resistance to change, fostering learning, and promoting collaboration.

INTEGRATING POKA-YOKE WITHIN THE LEAN-AI MODEL

Poka-yoke consists of operational, tactical, and strategic aspects for mistake-proofing organizations, systems, processes, and procedures as shown in Figure 7.2. The operational aspects of poka-yoke-AI deal with the day-to-day operational activities like mistake-proofing tasks. The operational aspects of poka-yoke-AI are the specific actions related to the daily tasks required to be completed without errors. For example, a simple poka-yoke-AI device can automatically help the operators place switches in the correct orientation for an error-free assembly.

The tactical aspects of poka-yoke-AI focus on the processes and procedures to ensure efficiency and consistency. For example, a poka-yoke device can help quality personnel verify that the electrical connections of the assembled switches are reliable, and meet the voltage requirements for the next step in the process. AI can

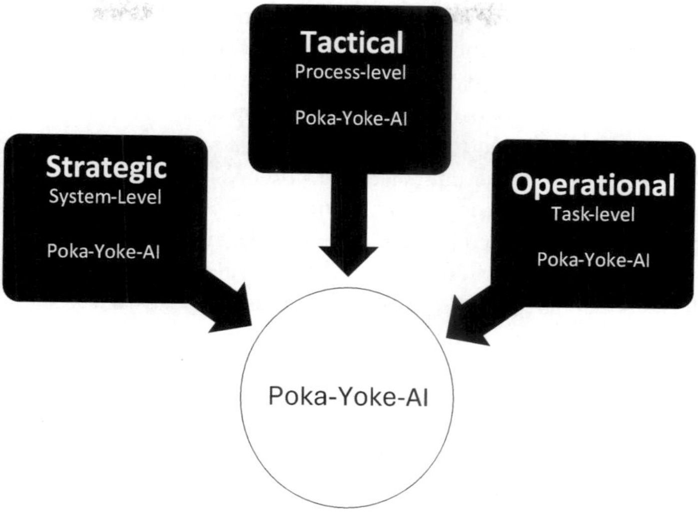

FIGURE 7.2 Poka-yoke-AI model.

enhance this by using historical data and predictive analysis to ensure that the process is completed accurately.

Finally, poka-yoke-AI has a strategic aspect. The strategic aspect of poka-yoke-AI is concerned with system-level safety and error-proofing. For example, applying poka-yoke to accounting software at the corporate level can prevent significant accounting errors for payroll, supplier accounts, and customer billing. AI allows the software to be self–enhancing taking the error-proofing capabilities to new levels.

The integrated poka-yoke-AI model provides specific tactical, operational, and strategic actions as well as upholds the two critical humanistic pillars of respect and humility, to ensure that poka-yoke systems meet the intended objective of zero defects and positively impact the organizational culture.

POKA-YOKE-AI DATA PRIVACY AND ETHICS CONSIDERATIONS

When integrating poka-yoke (error-proofing) with AI in organizational settings, data privacy and ethical considerations become essential, especially since AI-driven poka-yoke systems often require data about processes, human actions, and decision-making. Here are the key privacy and ethical elements to address for a responsible and effective deployment:

Privacy of Employee Performance Data: AI-powered poka-yoke may track employee interactions with equipment or workflows, potentially collecting performance data that could be personally identifiable. Ensure transparency about the types of data collected, informing employees about what is monitored and how the data will be used to support error-proofing rather than individual surveillance.

Minimizing Intrusiveness and Data Scope: Collect only the data essential for detecting and preventing errors, which could include information on machine usage, process parameters, or environmental factors. Avoid unnecessary collection of personal or sensitive employee data by limiting data scope to process-related metrics that serve poka-yoke goals, aligning with privacy principles of minimal data collection.

Anonymization, Aggregation, and Coding of Data: Where possible, anonymize, aggregate, or code data to reduce the risk of identifying specific employees, especially if AI-driven insights could reveal details about task performance. This approach allows AI to optimize poka-yoke systems while respecting individual privacy, focusing on process improvement rather than personal monitoring.

Bias and Fairness in AI-Driven Error Proofing: AI algorithms for poka-yoke should be designed and tested for fairness, as biases could lead to unfair blame or misidentification of errors, particularly in complex or diverse teams. Regularly audit AI outputs to ensure that the system does not unfairly flag certain groups of employees or teams as more error-prone due to biased data or assumptions.

Transparency and Explainability of AI Decisions: Employees should understand how AI-driven poka-yoke systems detect errors and provide feedback, which can help them trust and engage with the technology. Make the error-proofing mechanisms transparent by explaining how AI detects and prevents errors, so workers see AI as a support tool rather than a disciplinary measure.

Human Oversight and Empowerment: Maintain a human-in-the-loop model where employees can override or provide feedback on AI-generated error alerts to avoid dependency on AI and prevent erroneous decisions. Empower employees to be part of the AI-driven error-proofing process, enabling them to give input on AI recommendations and suggest improvements where necessary.

Protecting Against Excessive Surveillance: Avoid AI-driven poka-yoke solutions that intrusively monitor employees' every move under the guise of error-proofing, as this can create a stressful and distrustful work environment. Instead, focus on AI tools that enhance safety and efficiency without compromising worker autonomy, protecting against the misuse of AI for surveillance beyond error-proofing.

Ensuring Accountability in AI-Driven Error Detection: Assign accountability for the AI's error-proofing decisions, ensuring that employees are not unfairly blamed for errors flagged by AI. Implement mechanisms for review and appeal, allowing employees to contest incorrect AI-detected errors and promoting a fair and responsible deployment of poka-yoke systems.

Clear Communication on AI's Role in Process Improvement: Clearly communicate that AI-driven poka-yoke is designed to enhance process reliability and reduce human error rather than punish individuals. Reinforce that the objective is to create a safer, more efficient environment, aligning with organizational values of learning and continuous improvement rather than surveillance.

Balancing Efficiency with Ethical Workforce Implications: Evaluate the impact of AI-driven error-proofing on job roles and responsibilities, particularly if the AI's functionality automates aspects of human error prevention. Maintain ethical considerations by offering training or reskilling opportunities for employees who may be impacted by more automated or AI-led poka-yoke systems, ensuring that efficiency gains do not come at the expense of job security or employee morale.

Governance and Regular Ethical Review: Establish an ethics and governance team to oversee the use of AI in poka-yoke implementations, ensuring that data privacy, transparency, and fairness are continuously monitored. This governance body can review ethical concerns, employee feedback, and privacy risks, adjusting policies as needed to maintain trust and uphold organizational values.

By addressing these considerations, organizations can integrate poka-yoke with AI in a manner that respects employee privacy, enhances safety, and drives error reduction without compromising ethical standards. This careful balance between AI capabilities and human-centered practices supports sustainable, trust-driven innovation in error-proofing initiatives.

POKA-YOKE-AI INTEGRATION INDUSTRY EXAMPLES

Integrating poka-yoke with AI can significantly enhance error prevention across various industries as follows:

MANUFACTURING:

The manufacturing sector can benefit significantly from poka-yoke-AI in applications like quality control, assembly processes, predictive maintenance, and supply chain management. For example:

AI-enhanced vision systems can detect even minor errors in products on high-speed assembly lines like bottling operations, ensuring that only defect-free products advance to the next stage. Human eyesight and regular poka-yoke are incapable of such precision and accuracy in quality control applications.

AI-assisted systems can help operators complete complex processes with zero defects by providing real-time feedback and alerts of missed or incorrect steps.

AI can take predictive maintenance to the next level and prevent breakdowns by analyzing historical data from machinery, accurately predicting the component reliability, and time of its failure, and allowing the equipment to operate within specified parameters.

AI can help reduce the likelihood of errors in inventory levels and ensure timely delivery of products by predicting customer demand and optimizing supply chain operations.

Automotive:

The automotive sector can benefit significantly from poka-yoke-AI in applications like automotive manufacturing, and driver assistance systems. For example:

During the manufacturing of automobiles, AI can monitor assembly lines for errors, and ensure vehicles are assembled correctly, by detecting missed components, and incorrect part installations.

AI systems, like automatic braking systems, can be installed into vehicles to assist drivers and enhance safety by monitoring drivers' behavior and the surrounding environment, and providing alerts or taking corrective actions to avoid accidents.

Healthcare:

The healthcare sector can benefit significantly from poka-yoke-AI in applications like administration of medication, monitoring of patients, and surgeries. For example:

AI can eliminate human errors in the medication dispensing processes by monitoring and verifying the systems, ensuring that patients consistently receive the correct dosage of all the prescribed medications.

AI systems can help healthcare providers and doctors by continuously monitoring patients' vital signs and allowing immediate intervention by alerting them of abnormalities.

AI-assisted surgical robots can ensure error-free surgical procedures by providing real-time feedback that alerts surgeons before they make a critical mistake, thus saving lives.

Financial Services:

The financial sector can benefit significantly from poka-yoke-AI in applications like compliance, auditing, and fraud detection. For example:

AI systems can reduce the risk of human error in financial reporting by automatically reviewing transactions and financial records to ensure compliance with the required regulatory requirements.

AI can help prevent fraud by conducting real-time pattern analysis of transactions to detect and prevent fraudulent activities and safeguard the financial institution and customers' financial assets.

Retail:

The retail sector can benefit significantly from poka-yoke-AI in applications like inventory management, and point-of-sale systems. For example:

AI systems can prevent issues like overstocking or stockouts by tracking inventory levels and detecting discrepancies and variances between recorded and actual stocks.

AI can protect retailers and customers by real-time verification of transactions and altering fraudulent activities or errors during checkout.

LOGISTICS:

The logistics sector can benefit significantly from poka-yoke-AI in applications like automating warehousing operations and route optimization. For example:

AI-powered robots can increase the efficiency of warehouse operations by reducing human errors in tasks like picking, packing, sorting, and shipping,

AI can use real-time data to optimize delivery routes, prevent delays, and ensure timely deliveries.

TELECOMMUNICATIONS:

The telecommunications sector can benefit significantly from poka-yoke-AI in applications like network monitoring, and customer support. For example:

AI can continuously monitor network performance and ensure network reliability by detecting and correcting issues in real time before they impact users.

AI chatbots can help reduce response errors, ensure high customer satisfaction, and provide accurate and consistent customer support.

EDUCATION:

The education sector can benefit significantly from poka-yoke-AI in applications like automated grading and customized learning. For example:

AI can assist in consistent and error-free evaluations by accurately grading assignments and exams.

AI can reduce the likelihood of learning gaps for students, by tailoring learning plans and ensuring that educational content is personalized to their needs.

The aforementioned examples demonstrate that integrating poka-yoke with AI benefits various industries by enhancing error prevention and improving efficiency, productivity, safety, quality, and reducing costs.

POKA-YOKE-AI ADVANTAGES

Integrating poka-yoke and AI offers significant benefits, a few of these are:

Enhanced Real-Time Error Detection: Traditional poka-yoke mostly relies on visual, manual, or sensor-based error detection. An advantage of AI systems is that they are capable of real-time analysis of large amounts of data from multiple data sources. This enables more accurate and timely detection of errors, improving the effectiveness of error detection as compared to the traditional poka-yoke approach.

Predictive Capabilities: AI uses historical data to identify patterns indicative of future issues. Using this predictive capability AI can foresee failures before they occur. Proactively addressing these risks can assist organizations prevent defects, failures, and downtime, leading to improved productivity, cost savings, and a competitive advantage.

Self-Learning: An important benefit of AI is that it is self-learning. This can help the poka-yoke systems rapidly learn from their errors and continuously improve them to almost zero-defect levels. This iterative approach enables organizations to stay ahead of emerging risks and maintain the highest levels of accuracy, quality, and efficiency.

To summarize, integrating poka-yoke and AI can help organizations reach a higher accuracy of error detection and prevention, improve cycle times, increase process efficiencies, enhance product quality, reduce production costs, and ultimately drive competitive advantage.

CONCLUSION

Poka-yoke is a Japanese term and a Lean tool that refers to mistake-proofing or error-proofing a product, process, or system. The objective of error-proofing is zero defect. However, because human mistakes are inevitable, some major disadvantages of the capabilities of the current poka-yoke system, achieving zero defects is not practical.

The future of Lean requires a technology upgrade. Poka-yoke-AI can provide numerous advantages over the current poka-yoke and could bring us closer to achieving the objective of zero defects. It can enhance customer satisfaction and transform industries through real-time analysis and error detection, predicting errors, and self-learning capabilities, that has the potential to allow it to prevent errors at the source, and eliminate almost all defects.

EXERCISES

1. What are the current system and process limitations of your poka-yoke approaches, and how can you apply AI to overcome these limitations?

Poka-yoke Approach	System and Process Limitations	AI Approach to Overcome System and Process Limitations
1. Elimination		
2. Prevention		
3. Detection		
4. Correction		
5. Loss Control		

2. What are the current cultural limitations of your poka-yoke approach, and how can you apply AI to overcome these limitations?

Poka-yoke Approach	Cultural Limitations	AI Approach to Overcome Cultural Limitations
1. Elimination		
2. Prevention		
3. Detection		
4. Correction		
5. Loss Control		

3. What are some short- and long-term benefits to your department and organization for integrating AI into poka-yoke?

#	**Short-Term Benefits**	**Long-Term Benefits**
Example	*Increased accuracy*	*Improved product quality*
1		
2		
3		
4		
5		
6		

4. Reflect on the ethical implications of using AI in poka-yoke within your organization. Consider aspects such as employee privacy, job displacement, and data security. Develop strategies to mitigate the negative aspects of the workforce.

#	**Ethical Considerations**	**Strategies to Mitigate**
Example	*Job displacement fears*	• *Provide upskilling and reskilling programs* • *Clear communication about AI's role*
1		
2		
3		
4		
5		
6		

5. What are the technology considerations for integrating AI into your poka-yoke framework within your organization? Develop strategies for the necessary technology integration.

#	**Technology Necessary for AI**	**Strategy for Integration**
Example	*Real-time data analytics*	• *Develop robust data infrastructure* • *Train staff on data management*
1		
2		
3		
4		
5		
6		

REFERENCES

Hines, P., & Rich, N. (1997). The seven value stream mapping tools. *International Journal of Production Economics*, *45*(1–3), 39–58. https://doi.org/10.1016/S0925-5273(96)00023-5

Lazarević, M., Šarac, M., & Banjac, S. (2023). A systematic literature review of poka-yoke and novel approach to theoretical aspects. *Strojniski vestnik – Journal of Mechanical Engineering*, *69*(2), 123–135.

Lazarevic, M., et al. (2019, July-August). A systematic literature review of poka-yoke and novel approach to theoretical aspects. *Strojniski Vestnik – Journal of Mechanical Engineering*, *65*(44020), 454+. Gale Academic OneFile. link.gale.com/apps/doc/A628947514/AONE?u=anon~937edfec&sid=googleScholar&xid=30badbd3. Accessed 27 May 2024.

Rother, M., & Shook, J. (2003). *Learning to see: Value stream mapping to add value and eliminate Muda*. Lean Enterprise Institute.

Shingo, S. (1986). *Zero quality control: Source inspection and the poka-yoke system.* Productivity Press.

8 SMED-AI

SMED OVERVIEW

SMED is a Lean tool designed to reduce setup or changeover time, making these transitions a standard, routine operation with minimal impact on continuous flow. By implementing SMED, organizations benefit from reduced inventory, improved flexibility, increased capacity, and enhanced customer response. This chapter provides an overview of SMED and explores strategies for optimizing it through AI integration.

Integrating AI into SMED can significantly enhance its benefits by improving efficiency and effectiveness across its five main steps. AI-driven technologies enable faster changeovers, reduce downtime, and improve product quality, ultimately enhancing overall production efficiency. These improvements streamline operations while providing a competitive advantage, particularly in dynamic environments. To fully understand this integration, it is essential to first establish the foundational principles of SMED before exploring AI-driven enhancements.

SMED IN THE OFFICE

While SMED is traditionally associated with manufacturing, its principles can be effectively applied in office environments to optimize workflows and reduce downtime in administrative processes. Office setups often involve switching between tasks, accessing different systems, or reconfiguring workstations for various projects. Applying SMED in an office setting can minimize time wasted on inefficient task transitions, improve document management, and streamline workflows by identifying internal and external setup steps.

For example, in an office environment, internal setups may include gathering necessary files, logging into different software, or preparing reports, while external setups involve preloading templates, automating data entry, and standardizing documentation procedures. By systematically organizing these processes, businesses can reduce time spent on administrative tasks, improve employee productivity, and enhance responsiveness to customer needs. AI further enhances SMED in office settings by automating repetitive tasks, providing intelligent recommendations for workflow optimization, and predicting resource allocation for increased efficiency.

SMED OVERVIEW

Shigeo Shingo introduced SMED after observing die press and machine tool setup operations for 19 years (Cudney et al., 2013). In observing setup operations, Shingo determined that setups can be divided into two categories: internal and external (Emekdar et

DOI: 10.1201/9781003532972-8

al., 2023). Internal setups or steps are performed when the machine is not operating. For example, changing the fixture in the machine must be done when the machine is not running. External setups or steps are performed while the machine is running. For example, tools can be set for the following product while the machine is running.

Shingo also noted that converting internal setups to external setups can drastically reduce changeover times. The changeover time is calculated as the time from completing the last good piece of the current product to the first good piece of the next product (Cudney et al., 2023). Subsequently, changeover time includes changing materials, tools, fixtures, dies, and any inspection or testing necessary to ensure the next product meets the customer's expectations.

Reducing changeover time increases the time available to run production. Shingo stated that changeover times should be able to be reduced by 59/60ths. Ideally, changeovers should be instantaneous or as minimal as possible so that the continuous production flow is not affected. While that is the ultimate goal, SMED strives for a changeover time of less than 10 minutes, hence the term "single-minute."

The SMED methodology includes five main steps to enhance setup efficiency:

1. Determine the setup operation – Identify the area for improvement within the current setup and assess its impact on production flow.
2. Prepare for changeover – Ensure all necessary tools and materials are ready and operational before initiating the changeover.
3. Changeover internal steps – Execute the internal setup operations that require halting the machine.
4. Perform calibrations – Conduct precise measurements and calibrations to maintain quality.
5. Run trials and adjust – Fine-tune adjustments after initial production to ensure consistent output.

By implementing SMED in both manufacturing and office settings, organizations can optimize their operational efficiency and significantly reduce wasted time and resources.

STEP 1: DETERMINE THE SETUP OPERATION

This initial step involves identifying the area for improvement within the current setup. Instead of considering machines in isolation, the focus is on integrating setup time reduction into the entire production flow. The manufacturing process should smoothly transition from raw materials to finished products across processing, inspection, transport, and storage phases. To kickstart the evaluation, specialists often target the largest monument machine in the production facility to assess its upstream and downstream impacts, providing a foundation for process evaluation. The largest machine may or may not be the SMED to be analyzed, but it provides a useful starting point for process evaluation.

STEP 2: PREPARE FOR CHANGEOVER

Following the selection of the SMED evaluation candidate, the subsequent step entails ensuring all necessary tools and materials are ready and operational. This step is conducted externally to prepare for the changeover. This preparation phase plays a pivotal role in streamlining the setup process.

STEP 3: CHANGEOVER INTERNAL STEPS

After the preparatory phase, attention shifts to executing internal steps of the changeover process. This step involves removing parts and tools, post-processing a piece, and then installing the requisite components for the subsequent part. These tasks typically necessitate halting the machine; therefore, they are classified as internal setup operations.

STEP 4: PERFORM CALIBRATIONS

Accurate measurements and calibrations are performed during production operations to maintain precision and quality. While traditionally internal, these tasks can be adapted into external operations for efficiency.

STEP 5: RUN TRIALS AND ADJUST

The final stage of the changeover entails fine-tuning adjustments after machining the first piece. Leveraging precise measurements and calibrations from the previous step facilitates this process. In the traditional setup paradigm, this phase is considered an internal setup, as the changeover remains incomplete until satisfactory parts are produced.

ADVANTAGES OF SMED

There are numerous benefits to implementing SMED across various operational aspects. Organizations can reduce inventory, improve flexibility, increase capacity, and enhance customer service by reducing changeover time.

First, SMED streamlines process changes, significantly reducing the likelihood of errors by minimizing physical adjustments (Cudney et al., 2013). This enhanced efficiency extends to scheduling, where setup times are kept shorter than takt time, allowing for greater flexibility and agility in meeting production demands.

SMED also helps avoid the expense associated with excess inventory and increases overall capacity by minimizing downtime during changeovers. This reduction in setup duration also leads to less material wastage due to scrap, contributing to cost savings and improved resource utilization. Additionally, the methodology enhances product quality by minimizing variation between setups and reducing the risk of large-scale defects in inventory.

Further, SMED contributes to enhanced quality by minimizing defects arising from setup errors and eliminating the need for trial runs. The ability to switch between tasks swiftly enhances customer service by enabling companies to adapt quickly to changing customer needs. This agility in production also facilitates the creation of cost-effective products with reduced waste, particularly in smaller quantities.

Beyond efficiency gains, faster changeovers from SMED implementation boost productivity by minimizing downtime. Moreover, SMED fosters a safer work environment by simplifying setups, thereby reducing the risk of physical strain or injury.

Additionally, dovetailing with the principles of 5S, SMED helps optimize physical space utilization, promoting a more organized and clutter-free workspace that enhances overall efficiency.

LIMITATIONS OF SMED

While SMED offers numerous benefits, such as reducing setup times and improving operational efficiency, some limitations exist. SMED is highly effective for reducing changeover times in manufacturing processes; however, it may be less applicable in other industries where setup times are not a significant factor or do not involve machinery and equipment.

The first main limitation of SMED is that it can be challenging to implement SMED in complex environments, particularly in larger manufacturing environments with numerous machines and processes. These complex environments make coordinating and standardizing setup procedures challenging across multiple equipment and teams.

Another limitation to implementing SMED is process variability. SMED may be less effective in environments with significant variability in process specifications, materials, or production volumes. This additional complexity makes standardizing setup procedures challenging in such cases.

In addition, some additional investment is required when implementing SMED, such as training, equipment, and process reorganization. This investment can be a barrier, particularly for smaller organizations with limited resources.

SMED results in improved operational efficiency. Maintaining these gains requires adherence to standardized procedures and oversight. The improvements can slowly erode without regular monitoring and reinforcement.

SMED CULTURE

As with any methodology, there are limitations from a cultural perspective that must be considered. These limitations often stem from resistance to change, entrenched work practices, and differing attitudes toward process improvement.

When implementing SMED, employees may resist SMED due to the fear of increased workload, job insecurity, or discomfort with the new process. This resistance to change is often felt because employees feel their existing methods are sufficient; therefore, they do not see the need for change (Agustiady & Cudney, 2022).

SMED is one of the more complex Lean methodologies, which leads to another limitation around the lack of understanding. Training on SMED is critical for appropriate implementation. Insufficient training and education lead to employees and management not fully understanding the principles and benefits of SMED. This lack of understanding also leads to poor implementation and resistance.

INTEGRATING SMED AND AI

Integrating AI into SMED can significantly enhance the benefits of the methodology by improving efficiency and effectiveness across the five main steps. By integrating

AI technologies into SMED, organizations can achieve faster changeovers, reduce downtime, improve product quality, and ultimately enhance production efficiency. These improvements streamline operations while also providing a competitive advantage, particularly in dynamic environments.

Organizations should foster a culture of continuous improvement and technological adoption to leverage AI and SMED integration fully. Encouraging collaboration between employees and AI systems, promoting openness to change, and providing training and support can help employees adapt to new technologies and processes. By embedding these cultural values, organizations can ensure AI's smooth and effective integration into SMED processes.

Organizations can use AI to learn from operational data and refine setup processes continuously. Machine learning models can adapt and improve setup efficiency by identifying bottlenecks and recommending process optimizations. Additionally, AI-powered NLP can analyze and extract insights from maintenance logs, operator manuals, and other documentation related to setup operations. These insights help to identify recurring issues and suggest improvements based on textual analysis. Each SMED step can be enhanced through AI.

STEP 1: DETERMINE THE SETUP OPERATION

In the first step, AI algorithms can analyze production data and identify opportunities for setup time reduction. AI can detect patterns in machine downtime, production inefficiencies, and material flow disruptions that impact setup operations. By considering the entire production flow and using AI-driven analytics, organizations can pinpoint which setups to prioritize for improvement.

STEP 2: PREPARE FOR CHANGEOVER

Organizations can use AI to predict when machines will require maintenance or replacement parts based on real-time data analysis of equipment performance. This proactive approach ensures that all tools and materials needed for changeover are ready and operational, minimizing setup delays due to unexpected maintenance issues.

STEP 3: CHANGEOVER INTERNAL STEPS

Organizations can use AI-powered robotics to automate internal setup operations during changeovers, such as parts removal and installation. Advanced robotics can work collaboratively with operators, speeding up the process while maintaining safety and precision. AI-powered robotics can take over physically demanding and repetitive tasks, reducing the risk of injuries and strain for employees. By automating hazardous tasks, such as handling heavy components or working with dangerous machinery, AI helps create a safer work environment. This shift protects employees and fosters a culture of safety and well-being within the organization.

STEP 4: PERFORM CALIBRATIONS

Organizations can use AI for real-time monitoring and adjusting machine calibrations during production. AI algorithms can analyze sensor data to ensure precision and quality, automatically adjusting when deviations are detected. This analysis reduces setup time by integrating calibration tasks seamlessly into production cycles.

STEP 5: RUN TRIALS AND ADJUST

In the final step, AI simulations can predict optimal settings for new production runs based on historical data and current conditions. By simulating different scenarios, AI can recommend adjustments that minimize trial-and-error during setup and initial production phases, reducing the time needed to achieve stable production.

INTEGRATING SMED WITHIN THE LEAN-AI MODEL

Integrating AI within the SMED methodology enhances efficiency and effectiveness across setup operations. SMED-AI combines tactical, operational, and strategic elements to streamline setup processes and reduce changeover times, as shown in Figure 8.1. The tactical aspects involve utilizing AI-driven data analytics to identify setup optimization opportunities and ensure all necessary tools and materials are ready for changeovers. Operationally, AI enables automated setup procedures, such as parts removal and installation, improving precision and speed. Strategically, AI supports continuous improvement by analyzing real-time production data to refine setup techniques and maintain long-term efficiency gains. By integrating AI throughout SMED practices, manufacturers achieve faster changeovers, minimize downtime, and enhance overall production agility and competitiveness.

Integrating AI within the SMED methodology also fosters a culture of respect and humility, as emphasized in the Shingo model. Respect within SMED-AI involves valuing the contributions of every team member involved in setup optimization efforts. AI technologies can facilitate collaboration by providing data-driven insights that empower teams to make informed decisions collectively. Humility in SMED-AI entails acknowledging that there is always room for improvement and embracing continuous learning and adaptation. By applying AI to SMED principles with respect and humility, organizations can cultivate a supportive environment where innovation thrives, and individuals are encouraged to contribute their expertise toward achieving common goals of setup time reduction and operational excellence. This holistic approach improves technical processes and enhances workplace morale and teamwork, driving sustained improvements in manufacturing performance.

SMED-AI DATA PRIVACY AND ETHICS CONSIDERATIONS

The integration of AI into SMED introduces data privacy and ethical challenges that organizations must address to ensure responsible implementation. AI systems rely on vast amounts of data, including machine performance metrics, production schedules, and employee interactions with equipment. Safeguarding this data is critical to

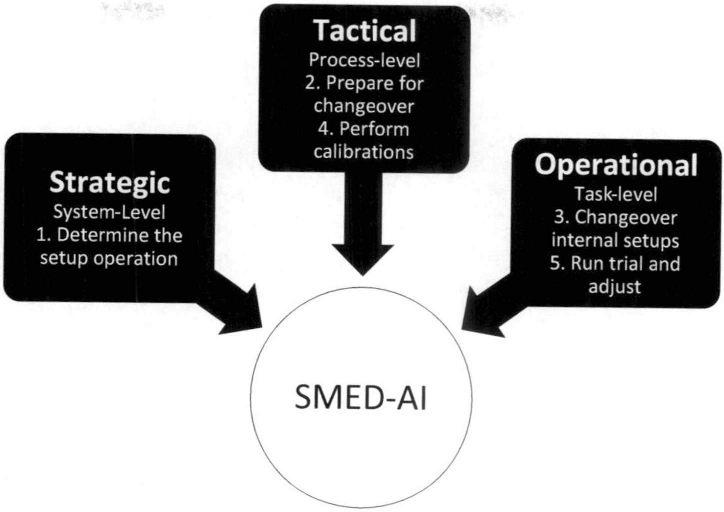

FIGURE 8.1 SMED-AI model.

maintaining compliance with data protection regulations and preventing unauthorized access.

One major ethical concern is employee surveillance. AI-powered monitoring systems can track worker efficiency, movements, and interactions with machines, potentially raising privacy concerns among employees. Organizations must implement AI in a transparent manner by ensuring that employees understand how data is being used and providing clear guidelines on privacy protections.

Bias in AI decision-making is another significant risk. If AI algorithms are trained on biased datasets, they may reinforce existing inefficiencies or inaccuracies in SMED processes. To mitigate this, organizations should use diverse datasets, regularly audit AI decisions, and incorporate human oversight into AI-driven recommendations.

Additionally, job displacement concerns arise when AI automates aspects of SMED that were previously performed by human workers. While AI can enhance efficiency, organizations must balance automation with workforce development by reskilling employees for new roles that complement AI-driven processes. Providing ongoing training ensures that employees remain engaged and that AI serves as a tool for augmentation rather than replacement.

Finally, cybersecurity risks must be managed when integrating AI into SMED. AI systems connected to industrial networks must be protected from cyber threats that could disrupt production or compromise sensitive data. Organizations should implement strong encryption, access controls, and continuous monitoring to safeguard AI-driven SMED processes.

By proactively addressing these privacy and ethical concerns, businesses can ensure that AI-enhanced SMED implementation aligns with organizational values, regulatory requirements, and employee well-being.

SMED-AI INTEGRATION INDUSTRY EXAMPLES

Integrating SMED and AI can significantly improve efficiency and productivity across various industries to enhance efficiency and effectiveness across setup operations.

In the automotive industry, SMED is used to rapidly changeover production lines for different car models or components. By integrating AI, manufacturers can utilize predictive maintenance and scheduling algorithms to minimize downtime and optimize production schedules (Inyang et al., 2023). AI can analyze production data to forecast the best changeover times, ensuring machinery runs at peak efficiency. This analysis reduces setup times and increases productivity, allowing manufacturers to respond more quickly to market demands and produce a wider variety of products without sacrificing efficiency.

In the healthcare industry, SMED is used to rapidly changeover operating rooms (ORs) between surgical procedures. By integrating AI, hospitals can utilize predictive maintenance and scheduling algorithms to minimize downtime and optimize OR schedules (Tlapa et al., 2022). AI can analyze historical and real-time data to forecast the best changeover times, ensuring equipment and rooms are prepared efficiently. This analysis reduces setup times and increases the number of surgeries that can be performed, allowing hospitals to respond more quickly to patient needs and improve overall care quality without sacrificing efficiency.

SMED techniques are applied in electronics manufacturing to switch production lines between electronic devices or circuit boards quickly. AI integration enhances this process using AI-driven vision systems that automate and optimize machinery setup and calibration (Koumas et al., 2021). These systems reduce setup times and minimize errors, ensuring consistent quality and faster production cycles. AI can also analyze production data to identify patterns and suggest improvements, further streamlining the changeover process and boosting overall efficiency.

Beverage bottling companies benefit from SMED by achieving fast changeovers of bottling lines for different beverage types and bottle sizes. Integrating AI into this process allows for the use of predictive algorithms to determine the optimal times for changeovers based on demand forecasting and real-time production data. This integration ensures that production lines are switched with minimal disruption, reducing downtime and increasing throughput (Jiménez-Delgado et al., 2024). AI also helps maintain quality and consistency across different product batches, enhancing overall operational efficiency.

In the pharmaceutical industry, SMED techniques enable efficient changeovers of production lines for different drugs or dosages. AI integration enhances this by optimizing production scheduling by considering demand, equipment availability, and maintenance schedules. AI algorithms can predict the most efficient changeover times, minimizing downtime and maximizing production output (Mehta et al., 2023). These predictions ensure a steady supply of various pharmaceutical products, improving the industry's ability to meet diverse patient needs and regulatory requirements.

SMED rapidly changes machinery in the textile industry for different clothing styles or sizes. AI integration improves this process by providing AI-driven scheduling and inventory management systems that align changeovers with demand forecasts (Dogan et al., 2018). These systems reduce idle times and excess inventory, ensuring that production lines always work on high-demand items. AI can also analyze trends and suggest optimal production schedules, helping manufacturers stay competitive in the fast-paced fashion industry.

Home appliance manufacturers use SMED to achieve quick setup changes for different appliance models or configurations. AI enhances this process through predictive analytics, which plans changeovers based on sales trends and production efficiency metrics (Abdullah et al., 2023). By minimizing downtime and optimizing production schedules, AI helps manufacturers increase productivity and meet market demands more effectively. This integration ensures that production lines can quickly adapt to new product launches and variations, maintaining a competitive edge.

In the aerospace industry, SMED is applied to efficiently changeover assembly lines for different aircraft models or components (Rojas et al., 2023). AI integration is crucial in this context, as AI-based predictive maintenance and scheduling ensure that changeovers are planned with minimal impact on production timelines. This integration enhances throughput and reduces downtime, allowing aerospace manufacturers to meet tight deadlines and maintain high quality and safety standards. AI can also analyze production data to identify improvements, further streamlining the assembly process.

SMED-AI ADVANTAGES

Integrating AI into SMED offers several significant advantages to industries in managing their production processes. One of the primary benefits is the drastic reduction in downtime during changeovers. AI-driven predictive maintenance and scheduling systems can analyze vast amounts of production data to forecast the optimal changeover times, ensuring machinery is utilized at peak efficiency. By anticipating when a machine will need maintenance or when a changeover will be most beneficial, AI minimizes unexpected disruptions, allowing production to flow more smoothly and consistently. This reduction in downtime directly translates to higher productivity and better resource utilization, essential for maintaining a competitive edge in fast-paced markets.

Another key advantage of integrating AI into SMED is improving quality control. AI systems, particularly those using machine learning and computer vision, can automate and optimize the setup and calibration of machinery during changeovers. This automation leads to more precise and consistent setups, reducing the likelihood of human error and ensuring higher product quality. AI can detect anomalies or deviations from the desired standards by continuously monitoring real-time production parameters, allowing immediate corrective actions. Real-time monitoring maintains high quality and reduces waste and rework, contributing to more sustainable manufacturing practices.

AI integration also brings significant improvements in flexibility and responsiveness to market demands. With AI-powered scheduling and inventory management systems, manufacturers can align their production changeovers with real-time demand forecasts. These systems ensure that production lines always work on high-demand items, reducing idle times and excess inventory. The ability to quickly adapt to market trends or changes in consumer preferences allows manufacturers to be more agile and responsive. In industries where product lifecycles are short and consumer preferences shift rapidly, this flexibility is invaluable for maintaining market relevance and customer satisfaction.

Furthermore, AI's ability to analyze and learn from production data provides valuable insights for continuous improvement. AI algorithms can identify patterns and suggest process optimizations that are not apparent through traditional analysis methods. By implementing these insights, manufacturers can continuously refine their SMED processes, leading to incremental improvements. This continuous learning and adaptation create a culture of ongoing improvement and innovation, essential for long-term success in any industry.

Finally, integrating AI into SMED supports better decision-making through data-driven insights. Managers and operators can rely on AI-generated reports and analytics to make informed decisions about production schedules, maintenance planning, and resource allocation. This data-driven approach reduces reliance on intuition or guesswork, leading to more accurate and strategic decisions. As a result, companies can optimize their operations holistically, balancing efficiency, quality, and responsiveness to achieve overall operational excellence.

CONCLUSION

SMED is the foundation of any Lean initiative. SMED is a systematic approach that promotes quick and efficient changeovers, which leads to enhanced productivity, flexibility, and efficiency. The SMED model emphasizes reducing setup times to streamline operations and minimize downtime. However, the major disadvantages of the current SMED model are that its success depends on human intervention and that it is a reactive model.

Integrating SMED with AI fosters a culture that embraces continuous improvement and safety. Organizations can significantly reduce setup times, enhance operational efficiency, and improve employee well-being, ultimately leading to better overall performance and competitiveness.

The future of Lean-AI requires a technology-driven foundation that we call SMED-AI. This AI-driven model will offer advantages. First, it will address the human dependence issue and make it predictive. SMED-AI helps organizations with data-driven decision-making, predictive maintenance, workflow optimization, employee training and engagement, and facility management. This integration ensures a more efficient, flexible, and responsive production process, aligning with the goals of Lean.

EXERCISES

1. What are the current system and process limitations of your SMED approach, and how can you apply AI to overcome these limitations?

SMED Step	System and Process Limitations	AI Approach to Overcome System and Process Limitations
Step 1: Determine the setup operation		
Step 2: Prepare for changeover		
Step 3: Changeover internal steps		
Step 4: Perform calibrations		
Step 5: Run trials and adjust		

2. What are the current cultural limitations of your SMED approach, and how can you apply AI to overcome these limitations?

SMED Step	Cultural Limitations	AI Approach to Overcome Cultural Limitations
Step 1: Determine the setup operation		
Step 2: Prepare for changeover		
Step 3: Changeover internal steps		
Step 4: Perform calibrations		
Step 5: Run trials and adjust		

3. What are some short- and long-term benefits to your department and organization for integrating AI into SMED?

#	Short-Term Benefits	Long-Term Benefits
Example	*Increased accuracy in setup*	*Improved product quality*
1		
2		
3		
4		
5		
6		

4. Reflect on the ethical implications of using AI in SMED within your organization. Consider aspects such as employee privacy, job displacement, and data security. Develop strategies to mitigate the negative aspects that impact the workforce.

#	Ethical Considerations	Strategies to Mitigate
Example	*Job displacement fears*	• *Provide upskilling and reskilling programs* • *Clear communication about AI's role*
1		
2		
3		
4		
5		
6		

5. What are the technology considerations for integrating AI into your SMED framework within your organization? Develop strategies for the necessary technology integration.

#	Technology Necessary for AI	Strategy for Integration
Example	*Real-time data analytics*	• *Develop robust data infrastructure* • *Train staff on data management*
1		
2		
3		
4		
5		
6		

REFERENCES

Abdullah, A., Saraswat, S., & Talib, F. (2023). Impact of smart, green, resilient, and lean manufacturing system on SMEs' performance: A data envelopment analysis (DEA) approach. *Sustainability*, *15*(2), 1379.

Agustiady, T., & Cudney, E. (2022). *Building a sustainable Lean culture: An implementation guide*. New York, NY: CRC Press.

Cudney, E., Furterer, S., & Dietrich, D. (2013). *Lean systems*. New York, NY: CRC Press.

Cudney, E., Furterer, S., Laux, C., & Hundal, G. (2023). *Lean sustainability: A pathway to a circular economy*. New York, NY: CRC Press.

Dogan, O., Cebeci, U., & Oksuz, M. K. (2018). An intelligent decision support system for SMED and its application in textile industry. In *Proceedings of the international conference on industrial engineering and operations management* (pp. 933–942). https://www.ieomsociety.org/paris2018/papers/169.pdf

Emekdar, E., Açikgöz-Tufan, H., Şahin, U. K., Kurşun Bahadir, S., Tuluk, B., & Şimşek, A. N. (2023). Process improvement and efficiency analysis using the single-minute exchange of dies method applied to the set-up and operation of screen-printing machines. *Coloration Technology*, *139*(2), 209–218.

Inyang, V., Kanakana, G. M., & Laseinde, O. T. (2023). Application of sustainable smart manufacturing technologies and toolkits in the automotive industry. *International Journal of Low-Carbon Technologies*, *18*, 412–422.

Jiménez-Delgado, G. I., Hernandez-Palma, H., Lucena-León Castro, B., & Ternera-Muñoz, Y. (2024, June). Lean philosophy "quick wins": A case study in a non-alcoholic beverage company. In *International conference on human-computer interaction* (pp. 279–294). Cham, Switzerland: Springer Nature.

Koumas, M., Dossou, P. E., & Didier, J. Y. (2021). Digital transformation of small and medium sized enterprises production manufacturing. *Journal of Software Engineering and Applications*, *14*(12), 607–630.

Mehta, A. K., Lanjewar, P., Murthy, D. S., Ghildiyal, P., Faldu, R., & Natrayan, L. (2023, December). AI & Lean management principles based pharmaceutical manufacturing processes. In *2023 10th IEEE Uttar Pradesh section international conference on electrical, electronics and computer engineering (UPCON)* (Vol. 10, pp. 1599–1604). IEEE.

Rojas, T., Mula, J., & Sanchis, R. (2023). Quantitative modelling approaches for lean manufacturing under uncertainty. *International Journal of Production Research*, 1–27.

Tlapa, D., Tortorella, G., Fogliatto, F., Kumar, M., Mac Cawley, A., Vassolo, R., . . . Baez-Lopez, Y. (2022). Effects of lean interventions supported by digital technologies on healthcare services: A systematic review. *International Journal of Environmental Research and Public Health*, *19*(15), 9018.

9 TPM-AI

TOTAL PRODUCTIVE MAINTENANCE OVERVIEW

Traditional manufacturing operates under the "we fix it" mentality, in which the maintenance department performs all maintenance activities. These activities are "firefighting" maintenance when a machine breaks down. The maintenance department performs some preventative maintenance; however, there is limited time for this because it is performed around the standard machine operating times. In the traditional manufacturing environment, the manufacturing department functions under the "we operate" and "run it "till it breaks" mentality. The operators generally do not perform any maintenance activities. Instead, the operators contact maintenance team when a machine breakdown occurs. In addition, the operators do not produce products during maintenance activities.

TPM is an innovative approach to equipment maintenance involving maintenance personnel and operators working in teams to eliminate equipment breakdowns and equipment-related defects (Cudney et al., 2013). There are three main goals of TPM:

1. Reduce unplanned equipment downtime
2. Eliminate barriers between departments
3. Reduce equipment-related defects

In addition, there are three main objectives, including:

1. Total employee involvement
2. Hands-on approach
3. Improve the organization's competitiveness

The TPM methodology of the four key phases is outlined in Figure 9.1 (Cudney et al., 2013). The methodology starts with returning equipment to almost new conditions. Next, the focus is on zero breakdowns through proper maintenance. The third phase focuses on consolidating information for future use. The final phase of TPM is zero defects.

Complete TPM implementation centers on autonomous maintenance, equipment improvement, maintenance prevention systems, and quality maintenance, as shown in Figure 9.2 (Cudney et al., 2023). Autonomous maintenance involves developing preventative maintenance practices. For equipment improvement, the equipment activities should focus on eliminating all breakdowns through physical equipment analysis (PEA) techniques to target zero breakdowns. The organization should develop an information system for maintenance prevention systems that consists of all TPM activities. Equipment-related decisions are then based on the maintenance

DOI: 10.1201/9781003532972-9

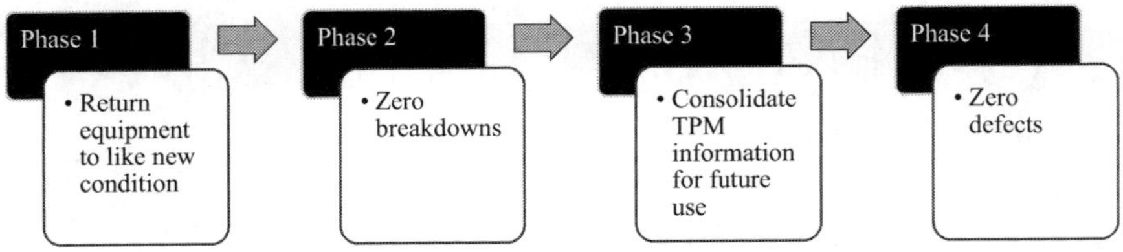

FIGURE 9.1 TPM phases.

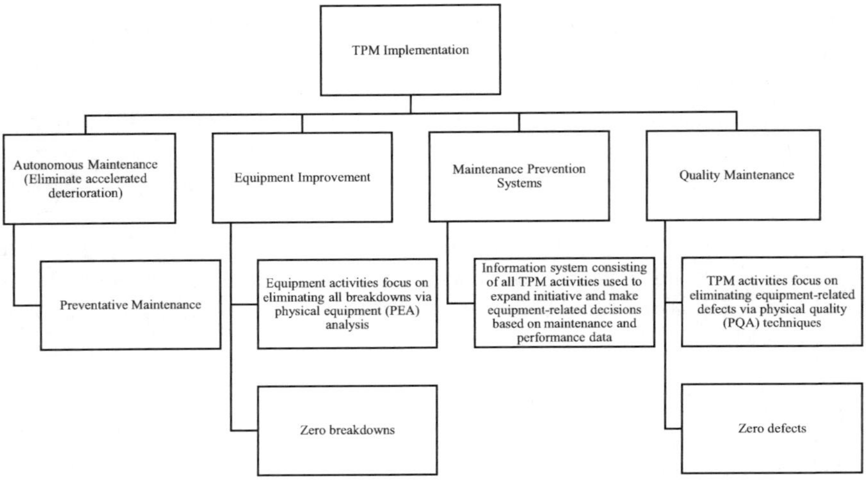

FIGURE 9.2 TPM implementation.

and performance data. Finally, quality maintenance eliminates equipment-related defects through physical quality analysis (PQA). The goal is zero defects.

Preventive maintenance (PM) is a time or usage-based method of maintaining equipment. Much like maintenance of oil changes in your automobile, maintenance activities are performed on equipment based on defined time or usage intervals to prevent equipment breakdowns from occurring. Examples include PM schedules and team activities.

Predictive maintenance is a situation-based method of maintaining equipment. Maintenance activities are performed on equipment based on visible signals or diagnostic techniques to prevent equipment breakdowns from occurring. Examples of predictive maintenance include vibration analysis, ultrasound, thermography, laser measuring, generator analysis, and oil analysis.

OEE measures the percentage of time a piece of equipment produces quality products. OEE measures the effect of the six big losses: breakdowns, setup, idling, minor stoppages, quality factors, and rework, as shown in Figure 9.3 (Agustiady & Cudney, 2023). These six big losses fall under the three categories of availability level, operating level, and quality level. The availability level consists of breakdowns and setups when the equipment is not available to run

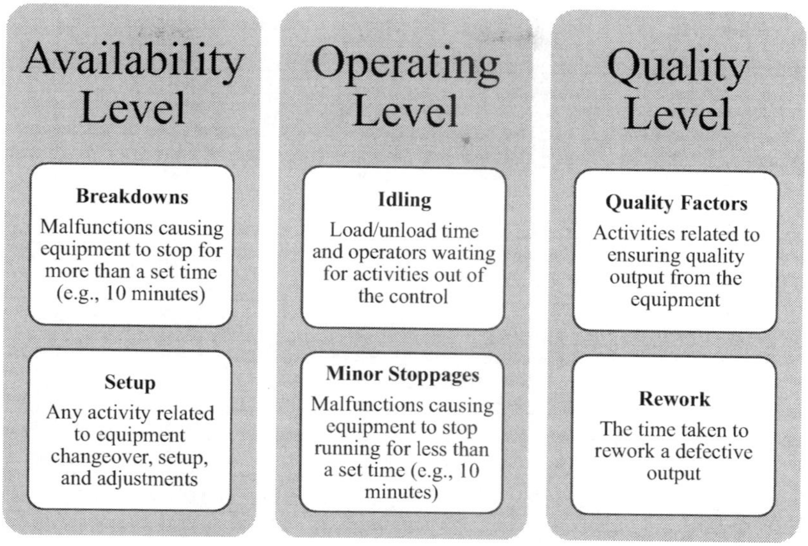

FIGURE 9.3 Major equipment losses.

production. The operating level consists of idling and minor stoppages, which occur when the equipment is not running; however, these are normal stoppages during the operation for activities such as unloading a part from a machine or malfunctions that are less than 10 minutes. The quality level consists of quality factors and rework, which relate to activities to ensure the product meets the customer's quality requirements or rework.

Overall equipment utilization and utilization are related concepts. Utilization or availability level captures the percentage of time lost due to breakdowns or setups. However, the OEE further defines the effectiveness of equipment by monitoring the operating and quality levels. The operating level considers the effect of load/unload time and minor stoppages, while the quality level considers inspecting output and rework. OEE indicates the actual time the equipment is producing a quality product. OEE takes utilization to a more detailed level.

OEE is important because it helps to prioritize improvement projects and reflect results. OEE also combines the equipment's utilization, operation, and quality aspects into one measure. OEE provides a measure for changes to capacity, productivity, and quality. Figure 9.4 provides the OEE calculation, and Figure 9.5 provides a sample calculation.

The initial TPM focus is on PM. The initial OEE calculations for equipment identify the current baseline. OEE will reflect improvements as the TPM initiative evolves. The initial focus of TPM is to return the equipment to like-new conditions and prevent further deterioration. Implementing PM schedules is the first step toward eliminating deterioration. A four-phased approach to PM schedules is presented in Figure 9.6.

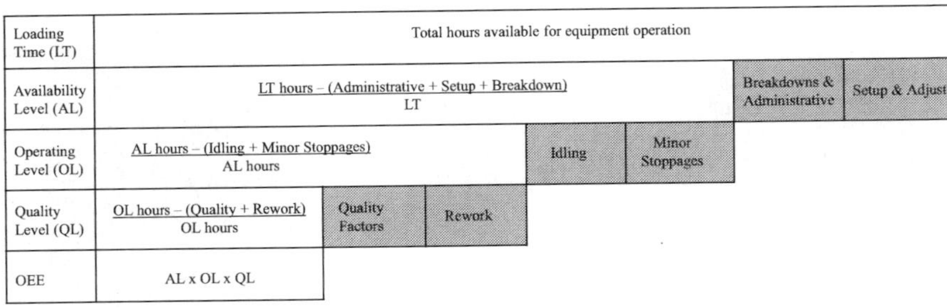

FIGURE 9.4 OEE calculation.

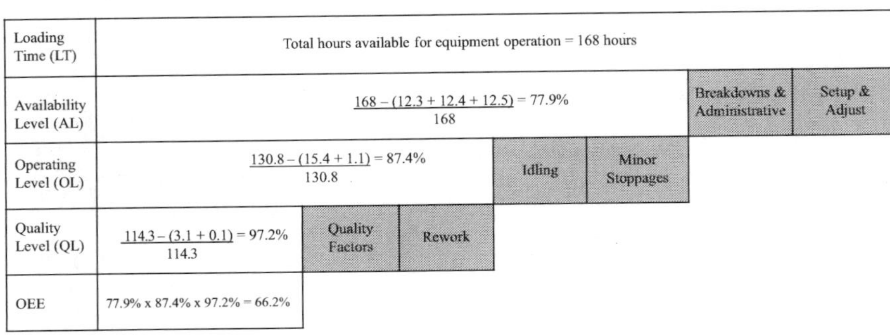

FIGURE 9.5 Sample OEE calculation.

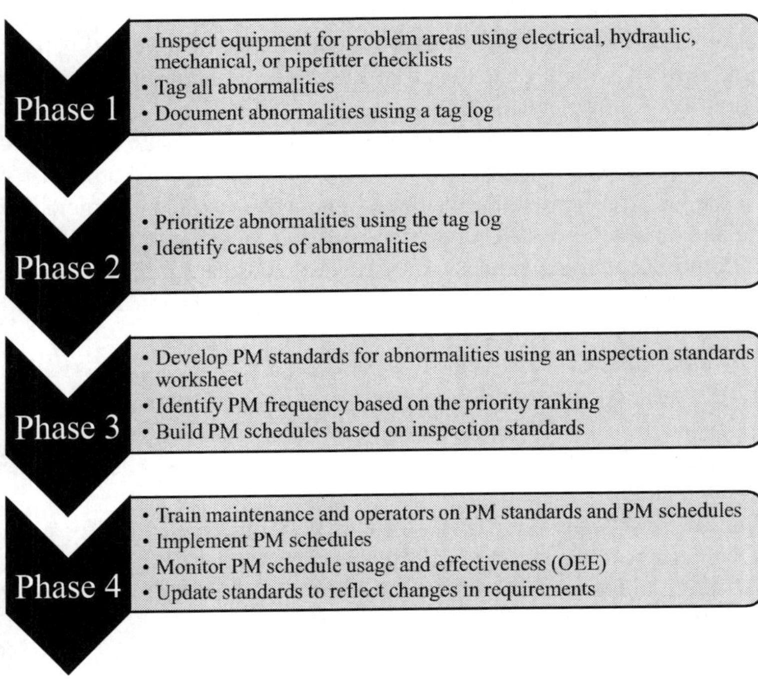

FIGURE 9.6 Four-phased preventive maintenance approach.

PHASE 1: INSPECTING TARGET EQUIPMENT

The first phase of PM schedule development begins with thoroughly inspecting the target equipment. The inspection involves utilizing customized checklists tailored to the specific equipment being inspected. These checklists examine all critical components and potential failure points systematically. During this inspection, any identified issues are tagged and documented in detail. This documentation includes the nature of the problem, the location, and the potential impact on equipment performance. Accurate and comprehensive tagging and documentation are crucial as they form the basis for subsequent analysis and maintenance planning.

PHASE 2: PRIORITIZING IDENTIFIED PROBLEMS

Once the inspection is complete and problems are documented, the next step is prioritizing these issues. Prioritization involves assessing the severity and impact of each problem to determine which ones require immediate attention. Commonly used tools in this phase include brainstorming sessions with maintenance teams, data collection from previous maintenance records, and leveraging the experience of operators and maintenance personnel. By combining these tools, organizations can identify the causes of the highest-priority problems. This phase may also involve root cause analysis to understand the underlying factors contributing to equipment issues. Prioritization ensures that maintenance efforts are focused on the most critical problems that could lead to significant downtime or equipment failure.

PHASE 3: DEVELOPING INSPECTION STANDARDS AND BUILDING PM SCHEDULES

The third phase focuses on developing inspection standards and building PM schedules. Inspection standards are written protocols outlining the procedures for conducting effective PM. These standards should be clear, concise, and tailored to the specific equipment and maintenance needs. They ensure consistency and accuracy in the maintenance process. Based on these standards, PM schedules are created to outline the timing and frequency of maintenance activities. These schedules should consider the equipment's operational patterns, historical performance data, and the results from the initial inspection and prioritization phases. Effective PM schedules balance the need for thorough maintenance to minimize operational disruption.

PHASE 4: TRAINING, IMPLEMENTING PM SCHEDULES, AND MONITORING

The final phase involves delivering effective training, implementing the developed PM schedules, and establishing monitoring mechanisms. Training is crucial to ensure that both operators and maintenance personnel understand and can effectively carry out the PM activities. This training should cover the specific inspection standards, the use of checklists, and the importance of accurate documentation. Once training is complete, the PM schedules are implemented, and maintenance activities begin according to the predefined timelines. Continuous monitoring is imperative to ensure the effectiveness of these activities. Continuous monitoring involves tracking equipment performance, using metrics such as OEE to measure improvements and identify areas for further optimization. Regular reviews and adjustments to the PM schedules based on monitoring data help maintain the equipment in optimal condition and ensure the sustainability of the TPM initiative.

ADVANTAGES OF TPM

Implementing TPM offers several significant advantages to organizations. TPM is a proactive approach focusing on maximizing equipment effectiveness and minimizing downtime through comprehensive maintenance strategies. One key advantage of TPM is its ability to enhance OEE. TPM reduces unplanned downtime and breakdowns by systematically addressing equipment reliability, maintenance, and operator involvement. This results in increased uptime for machines, directly translating into higher production output and improved productivity levels.

TPM also fosters a culture of continuous improvement and employee engagement. Through TPM initiatives, employees are empowered to take ownership of equipment maintenance and operation. This involvement enhances their skills and increases their sense of responsibility toward equipment care and operational efficiency. Engaged employees are more likely to identify potential issues early, perform routine maintenance tasks effectively, and contribute ideas for further process improvements, thereby driving overall operational excellence.

Another advantage of TPM is its contribution to reducing operational costs. By preventing breakdowns and addressing equipment issues proactively, TPM minimizes the need for costly repairs and emergency maintenance. Planned maintenance schedules and predictive maintenance techniques, supported by AI and data analytics, help optimize maintenance activities, extend equipment life cycles, and reduce spare parts inventory. This proactive maintenance approach leads to significant cost savings over time, contributing to the organization's bottom line.

Further, TPM supports quality improvement efforts within operations. Well-maintained equipment operates more consistently and reliably, resulting in fewer defects and rework. This consistency in production processes helps maintain product quality standards and enhances customer satisfaction. TPM's focus on PM and standardized work procedures ensures that production runs smoothly, meeting internal quality benchmarks and external customer expectations.

LIMITATIONS OF TPM

TPM offers numerous benefits but also faces several limitations that organizations must consider when implementing and sustaining this approach. One significant limitation is the required initial investment and resource allocation. Typically, implementing TPM involves significant upfront costs for training employees, acquiring new technologies for predictive maintenance, and establishing new processes and procedures. Therefore, it may be challenging for small and medium-sized enterprises (SMEs) or organizations with limited financial resources to justify these initial investments, especially if they do not have immediate visibility into the long-term benefits.

In addition, TPM implementation can be complex and time-consuming. It involves planning, data collection, and analysis to establish baseline equipment conditions and develop effective maintenance strategies. Organizations may struggle with integrating TPM with existing maintenance practices or other Lean initiatives, leading to potential conflicts or inefficiencies. Additionally, sustaining TPM requires ongoing commitment and dedication from all levels of the organization, as well as continuous monitoring and adjustment of maintenance activities based on evolving equipment needs and operational priorities.

Another limitation is the reliance on skilled maintenance personnel. Effective TPM implementation relies on having trained and skilled maintenance technicians capable of performing proactive and predictive maintenance tasks. The shortage of qualified personnel or high turnover rates in maintenance departments can pose challenges to sustaining TPM practices over the long term. Moreover, as technologies evolve, maintaining up-to-date knowledge and skills in new predictive maintenance techniques and AI applications becomes crucial, requiring ongoing investment in training and development.

Lastly, while TPM aims to minimize equipment downtime and improve OEE, it may not entirely eliminate unexpected breakdowns or disruptions in production. External factors such as supplier issues, environmental conditions, or unforeseen operational challenges can still impact equipment reliability and performance despite proactive maintenance efforts. Organizations must remain adaptable and resilient in their maintenance and operational management approach to address these unpredictable factors effectively.

CULTURE

A significant cultural limitation of TPM is the challenge of fostering a mindset shift toward proactive maintenance and continuous improvement across all levels of the organization. In traditional manufacturing or operational environments, where reactive maintenance practices have long been the norm, transitioning to TPM requires overcoming deep-seated habits and beliefs. Employees accustomed to addressing problems only when they arise may resist the proactive approach of TPM, viewing it as unnecessary or disruptive to established routines. This resistance can manifest as skepticism toward new procedures, reluctance to adopt unfamiliar technologies or methodologies, and a preference for the status quo.

Moreover, achieving buy-in from middle management and frontline employees is essential for the successful implementation of TPM. Middle managers often play a pivotal role in translating strategic goals into actionable plans and motivating their teams toward achieving those objectives. If middle management does not fully support TPM initiatives or fails to communicate its importance effectively to their teams, it can lead to lukewarm participation or outright employee resistance. This lack of alignment can undermine the cultural transformation necessary for TPM to thrive and sustain over the long term.

Inadequate leadership support is another cultural barrier that can impede TPM implementation. Effective leadership is essential for setting a clear vision, providing resources and guidance, and fostering a supportive environment for change. Leaders must champion TPM initiatives, communicate its benefits persuasively, and allocate sufficient resources for training, technology investments, and process improvements. Without strong leadership commitment, employees may perceive TPM as just another passing initiative or fail to see its relevance to their daily roles and responsibilities.

Another cultural limitation of TPM is the misconception that its principles and methodologies exclusively apply to manufacturing settings. This mindset overlooks the potential benefits of TPM in other industries, such as healthcare and service environments, where equipment reliability and operational efficiency are equally critical. For example, medical equipment is crucial in patient care and treatment outcomes in healthcare settings. Implementing TPM can ensure that medical devices are maintained in optimal condition, minimizing breakdowns and enhancing patient safety. However, healthcare professionals may perceive TPM as a concept tailored only for industrial machinery, overlooking its relevance to healthcare equipment maintenance. This limited perspective can hinder the adoption of TPM practices in healthcare facilities, leading to missed opportunities for improving equipment reliability, reducing downtime, and optimizing healthcare delivery processes.

Similarly, equipment reliability directly impacts service quality and customer satisfaction in service environments such as hospitality, transportation, or IT services. For instance, malfunctioning heating, ventilation, and air conditioning (HVAC) systems or elevators can disrupt guest comfort and experience in a hotel setting. Adopting TPM principles can help service providers implement proactive maintenance strategies to prevent equipment failures, reduce service interruptions, and maintain high service delivery standards. However, the belief that TPM is confined to manufacturing may prevent industry service professionals from exploring these methodologies and reaping their operational benefits.

Addressing these cultural limitations requires proactive efforts to build awareness, educate employees about the benefits of TPM, and involve them in the change process from the outset. Organizations in healthcare and service sectors must broaden their understanding of TPM beyond its traditional manufacturing context. Comprehensive training programs, workshops, and hands-on demonstrations can help employees understand how TPM contributes to their job satisfaction, operational efficiency, and the organization's overall success. Creating forums for open communication, feedback, and recognition of early successes can also foster a positive cultural shift toward embracing TPM principles and practices.

Finally, adapting TPM frameworks to suit the unique characteristics and requirements of healthcare and service environments is crucial. Adapting the TQM framework may involve tailoring maintenance schedules, performance metrics, and training programs to align with specific regulatory standards, patient care protocols, and service-level agreements. By customizing TPM implementation strategies to address sector-specific challenges and priorities, organizations can effectively overcome the misconception that TPM is limited to manufacturing and realize its full potential in enhancing equipment reliability and service delivery excellence.

INTEGRATING TPM AND AI

Integrating AI into the four phases of TPM can significantly enhance efficiency and the resulting benefits.

PHASE 1: INSPECTING TARGET EQUIPMENT

In the first phase, inspecting target equipment, AI can automate data collection and analysis. AI-powered IoT sensors continuously monitor equipment health, capturing real-time data from equipment such as temperature, vibration, and pressure. Automated data collection and analysis enable the detection of anomalies that may indicate potential issues. Furthermore, machine learning algorithms can analyze both historical and real-time data to identify patterns and predict equipment failures. This predictive capability allows maintenance teams to tag and document problems before they lead to breakdowns. AI can also generate dynamic checklists tailored to each piece of equipment's specific condition and history, ensuring thorough and targeted inspections.

PHASE 2: PRIORITIZING IDENTIFIED PROBLEMS

In the second phase, prioritizing identified problems, AI assists by assessing the criticality and impact of issues. AI algorithms can determine the severity and potential consequences of each problem, helping maintenance teams prioritize tasks based on risk levels and ensuring that the most critical issues are addressed first. AI-driven tools can analyze root causes by sifting through maintenance logs and historical data to accurately pinpoint underlying causes of high-priority problems using machine learning and NLP. Additionally, AI can synthesize data from various sources, including operator feedback and sensor data, providing a comprehensive view of equipment health and maintenance priorities.

PHASE 3: DEVELOPING INSPECTION STANDARDS AND BUILDING PM SCHEDULES

During the third phase, developing inspection standards and building PM schedules, AI enhances the process by analyzing best practices and historical data to create optimal inspection standards, which are continuously updated based on new insights. AI algorithms can develop dynamic PM schedules that adapt to the specific needs of each piece of equipment based on predictive maintenance models that forecast maintenance needs according to equipment usage patterns and health indicators. AI can also optimize maintenance procedures by identifying the most effective methods and tools for each task, improving the efficiency and effectiveness of PM activities.

PHASE 4: TRAINING, IMPLEMENTING PM SCHEDULES, AND MONITORING

In the final phase, which involves delivering training, implementing PM schedules, and monitoring and adjusting, AI supports the process through customized training modules for operators and maintenance personnel. AI-driven training platforms can employ virtual reality (VR) and augmented reality (AR) to simulate maintenance scenarios, enhancing hands-on learning. AI-powered systems continuously monitor equipment performance, providing real-time alerts and recommendations for maintenance activities, ensuring adherence to PM schedules and necessary adjustments. AI analyzes OEE and other performance metrics to evaluate the effectiveness of PM activities, identifying trends and areas for improvement to make data-driven adjustments. Additionally, AI systems create feedback loops that continuously gather data from maintenance activities and operational performance, refining PM schedules and training programs to ensure they remain effective and relevant.

By integrating AI into the four phases of TPM, organizations can significantly enhance their PM practices, improve equipment reliability, and achieve greater overall efficiency. This integration promotes a proactive maintenance culture, reducing unplanned downtime and maximizing the operational lifespan of equipment.

INTEGRATING TPM WITHIN THE LEAN-AI MODEL

Integrating AI within the TPM methodology significantly enhances efficiency and effectiveness across maintenance operations. AI-TPM combines tactical, operational, and strategic elements to streamline maintenance processes and reduce equipment downtime, as shown in Figure 9.7.

On a tactical level, AI-driven data analytics are crucial in identifying maintenance optimization opportunities. These systems analyze equipment performance data to predict potential failures and maintenance needs, enabling organizations to prepare the necessary tools and materials in advance. This proactive approach ensures that

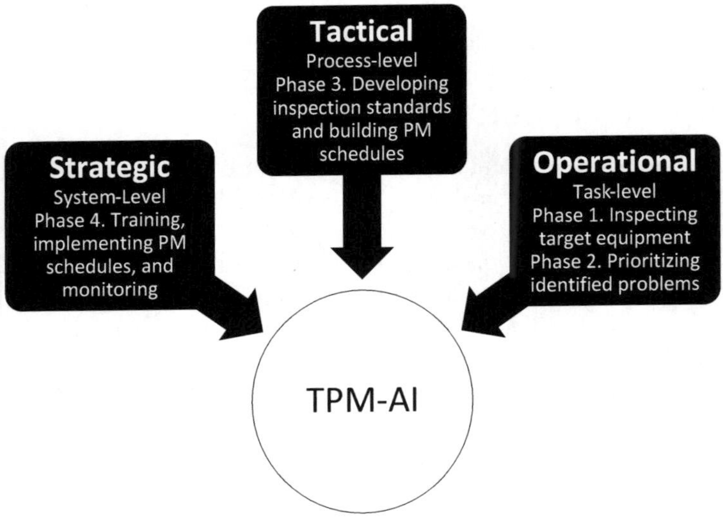

FIGURE 9.7 TPM-AI model.

maintenance tasks are carried out efficiently, minimizing unexpected breakdowns and interruptions.

Operationally, AI facilitates automated maintenance procedures, such as diagnostics, inspections, and minor repairs. Advanced robotics and AI-powered systems can perform these tasks with high precision and speed, reducing the burden on human operators and minimizing downtime. This automation improves maintenance accuracy and allows employees to focus on more complex and value-added activities, enhancing overall productivity.

Strategically, AI supports continuous improvement by analyzing real-time and historical maintenance data to refine maintenance techniques and strategies. This analysis helps identify patterns and trends, allowing organizations to optimize maintenance schedules, extend equipment lifespan, and maintain long-term efficiency gains. By integrating AI into their TPM practices, organizations can significantly reduce equipment downtime, enhance maintenance accuracy, and improve overall operational agility and competitiveness.

Integrating AI within the TPM methodology also fosters a culture of respect and humility, as emphasized in the Shingo model. Respect in AI-TPM involves valuing the contributions of every team member involved in maintenance optimization efforts. AI technologies facilitate collaboration by providing data-driven insights that empower teams to make informed decisions collectively. This collaborative approach ensures that the expertise and insights of all team members are considered and utilized effectively, fostering a sense of ownership and involvement.

Humility in AI-TPM means acknowledging that there is always room for improvement and embracing continuous learning and adaptation. AI systems help organizations identify areas for improvement and encourage a mindset of ongoing development. By applying AI to TPM principles with respect and humility, organizations can cultivate a supportive environment where innovation thrives, and individuals are encouraged to contribute their expertise toward achieving common goals of maintenance efficiency and operational excellence.

This holistic approach improves technical processes and enhances workplace morale and teamwork, driving sustained improvements in equipment reliability and overall manufacturing performance. Integrating AI into TPM practices enables organizations to leverage advanced technologies for better maintenance outcomes while fostering a positive and collaborative organizational culture.

TPM-AI DATA PRIVACY AND ETHICS CONSIDERATIONS

As AI becomes an integral part of TPM, organizations must address several data privacy and ethical concerns. AI-driven predictive maintenance relies on vast amounts of operational data, often collected through sensors, IoT devices, and machine learning algorithms. While these technologies offer significant advantages in improving equipment reliability and reducing downtime, they also introduce risks related to data security, employee privacy, and ethical decision-making.

One of the primary concerns with AI-powered TPM systems is employee surveillance and privacy. AI-driven maintenance platforms continuously collect and analyze data on equipment usage, work patterns, and operator behavior. While this data is crucial for predictive maintenance, it may inadvertently lead to increased monitoring of employees. Organizations must ensure transparency in data collection policies, clearly communicating how employee-related data is used and safeguarding against unauthorized surveillance.

Another ethical challenge is bias in AI decision-making. AI models rely on historical data to predict maintenance needs and identify equipment failures. However, if the training data contains inherent biases or incomplete information, AI-driven recommendations may favor certain maintenance strategies over others, leading to suboptimal or unfair decision-making. Organizations should implement robust validation mechanisms to detect and mitigate biases in AI algorithms, ensuring fair and accurate maintenance practices.

Cybersecurity risks are also a critical factor in AI-driven TPM. As maintenance systems become more interconnected, they are increasingly vulnerable to cyber threats, such as data breaches, ransomware attacks, or unauthorized access to predictive maintenance insights. Organizations must implement strong cybersecurity measures, including encryption, secure access controls, and regular system audits, to protect sensitive maintenance data and prevent potential disruptions.

Another key consideration is data ownership and accountability. As AI-powered TPM systems rely on data from multiple sources – including equipment manufacturers, third-party maintenance providers, and internal operations – organizations must establish clear policies on data ownership and usage rights. Defining who has access to AI-driven maintenance insights and how these insights are used is essential to maintaining ethical standards and avoiding conflicts over proprietary information.

Finally, workforce displacement and job security remain major ethical concerns. While AI enhances maintenance efficiency and reduces downtime, it may also automate tasks traditionally performed by maintenance technicians. Organizations should proactively address this challenge by investing in reskilling and upskilling programs, ensuring that employees can transition to roles that complement AI-driven TPM rather than being replaced by it. Encouraging human-AI collaboration rather than full automation can help create a balanced, ethical approach to AI integration.

To successfully implement AI in TPM while addressing these ethical considerations, organizations should develop AI governance frameworks that outline data privacy policies, ethical AI guidelines, and cybersecurity best practices. By fostering transparency, fairness, and accountability in AI-driven TPM, companies can harness the benefits of predictive maintenance while maintaining ethical integrity and workforce trust.

By incorporating data privacy and ethics considerations into AI-driven TPM, organizations can ensure responsible AI deployment while maximizing operational efficiency and equipment reliability.

TPM-AI INTEGRATION INDUSTRY EXAMPLES

Integrating TPM and AI offers transformative benefits across various industries. Organizations can improve equipment reliability, efficiency, and performance by combining TPM's proactive, team-based approach with AI's predictive and analytical capabilities. This integration allows for real-time monitoring, data-driven decision-making, and predictive maintenance, which together minimize unplanned downtimes, reduce maintenance costs, and extend the lifespan of critical machinery and equipment. The following paragraphs provide industry-specific examples illustrating how implementing TPM and AI can drive significant operational improvements, enhance productivity, and contribute to overall organizational success.

In the manufacturing industry, implementing TPM and AI has revolutionized the maintenance of machinery and production lines. AI-driven predictive maintenance systems continuously monitor equipment conditions using sensors and IoT devices (Siray, 2018; Antosz et al., 2020). For instance, in an automotive manufacturing plant, AI algorithms analyze data from machine vibrations (Mohan et al., 2021), temperature fluctuations (Alseiari & Farrell, 2021), and operational sounds (Çınar et al., 2020) to predict potential failures before they occur. This proactive approach minimizes unplanned downtimes, reduces maintenance costs, and extends the lifespan of critical machinery. By integrating AI into TPM, manufacturers can ensure that their equipment is always in optimal working condition, increasing production efficiency and reducing operational risks.

In the healthcare sector, implementing TPM combined with AI has significantly improved the reliability and performance of medical equipment. Hospitals and clinics use AI-powered predictive maintenance systems to monitor the functionality of MRI machines (Barretiri, 2021), ventilators (Ruhiyat et al., 2022), and other critical medical devices (Pokvić et al., 2023). For example, AI algorithms analyze data from various sensors to detect anomalies in an MRI machine's performance, allowing maintenance teams to address issues before they lead to equipment failure (Barretiri, 2021). This approach ensures the availability of essential medical devices and enhances patient safety and care quality. By integrating TPM and AI, healthcare facilities can maintain high operational efficiency and equipment reliability standards, ultimately improving patient outcomes.

In the energy sector, particularly in power plants and renewable energy facilities, TPM and AI integration aids in maintaining the efficiency and reliability of equipment. AI-powered predictive maintenance tools monitor wind turbines (Viharos et al., 2010), generators (Ćelić et al., 2024), and solar panels (Kahraman et al., 2018). For instance, in a wind farm, AI systems analyze data from sensors on wind turbines to predict potential mechanical failures, optimize maintenance schedules, and reduce downtime (Tjernberg, 2023). This proactive maintenance approach helps maximize energy production, reduce operational costs, and extend the lifespan of expensive equipment. By leveraging AI within TPM, energy companies can ensure a steady and reliable supply of energy, contributing to overall sustainability and efficiency.

In the food and beverage industry, implementing TPM and AI has enhanced the maintenance and operation of production lines (Sukma et al., 2021). AI-driven predictive maintenance systems monitor packaging machines (Chen & Voigt, 2020), conveyor belts (Uzoigwe, 2024), and refrigeration units (Chen & Voigt, 2020). For example, in a beverage bottling plant, AI algorithms analyze sensor data to predict when a conveyor belt might fail or a refrigeration unit might require maintenance (Alatrista-Corrales et al., 2021). This proactive approach minimizes production stoppages, ensures consistent product quality, and reduces maintenance costs. By integrating TPM and AI, food and beverage companies can maintain high production standards, reduce waste, and improve overall operational efficiency.

The aerospace industry benefits significantly from integrating TPM and AI into maintaining aircraft and related equipment (Korchagin et al., 2022). AI-powered predictive maintenance systems monitor aircraft engines (Pundir et al., 2022), landing gear (Rabathi & Hasbullah, 2023), and other critical components. For instance, airlines use AI algorithms to analyze data from aircraft sensors to predict potential failures and schedule maintenance activities accordingly (Fei et al., 2020). This approach ensures the safety and reliability of aircraft, minimizes flight delays and cancellations, and optimizes maintenance costs. By implementing TPM and AI, the aerospace industry can achieve higher safety, reliability, and operational efficiency levels, ultimately enhancing passenger satisfaction and trust.

In each of these industries, integrating TPM and AI leads to a proactive and efficient maintenance approach, ensuring that equipment is always in optimal condition, reducing downtime, and improving overall operational performance.

TPM-AI ADVANTAGES

Integrating TPM and AI offers many advantages that can significantly enhance operational efficiency, reliability, and overall productivity across various industries. One of the primary benefits is the reduction of unplanned downtimes. AI-driven predictive maintenance systems continuously monitor equipment conditions and analyze data to predict potential failures before they occur. This proactive approach allows maintenance teams to address issues in advance, minimizing unexpected breakdowns and ensuring that machinery operates smoothly and without interruption. Reduced downtime translates into higher production rates, better utilization of resources, and increased profitability.

Another significant advantage of combining TPM and AI is the enhancement of equipment lifespan and reliability. By utilizing AI to monitor and analyze the performance of machinery, organizations can implement more effective maintenance strategies that prevent wear and tear. Predictive maintenance based on AI insights ensures that equipment receives timely care, reducing the risk of severe damage and extending its operational life. PM not only leads to cost savings by delaying the need for expensive replacements but also improves the reliability and efficiency of the production process. Consistent equipment performance ensures that production targets are met and quality standards are maintained, which is crucial for customer satisfaction and competitive advantage.

Integrating TPM and AI also promotes a culture of continuous improvement and total employee involvement. AI provides data-driven insights that empower employees at all levels to make informed decisions regarding maintenance and operations. Organizations can foster ownership and accountability by involving operators and maintenance personnel in the maintenance process and equipping them with AI tools. This collaborative environment encourages employees to contribute their expertise and ideas to improve maintenance practices and overall operational efficiency. The hands-on approach promoted by TPM, combined with the advanced capabilities of AI, leads to a more engaged and motivated workforce, which is essential for sustaining long-term improvements and achieving organizational goals.

Further, implementing TPM and AI facilitates better resource management and cost optimization. AI's ability to analyze vast amounts of data and provide actionable insights enables organizations to optimize their maintenance schedules, allocate resources more effectively, and reduce unnecessary expenditures. For example, AI can identify patterns in equipment usage and maintenance needs, allowing for more precise planning and allocation of maintenance tasks. Proper maintenance planning ensures that resources such as spare parts, labor, and time are used efficiently, reducing waste and lowering overall maintenance costs. Improved resource management also contributes to the organization's sustainability goals by minimizing the environmental impact associated with excessive resource consumption and waste generation.

Lastly, the combination of TPM and AI enhances overall operational safety. AI-powered systems can detect anomalies and potential hazards in real time, alerting maintenance teams to address issues before they escalate into safety risks. This proactive approach to maintenance helps create a safer working environment for employees by preventing accidents and ensuring that equipment operates within safe parameters. Enhanced safety protects employees from harm and reduces the likelihood of costly incidents and legal liabilities. By prioritizing safety through TPM and AI integration, organizations can build a strong safety culture that promotes employee well-being and reinforces their commitment to operational excellence.

In summary, integrating TPM and AI offers numerous advantages, including reduced downtimes, extended equipment lifespan, enhanced reliability, a culture of continuous improvement, better resource management, cost optimization, and improved operational safety. These benefits contribute to higher productivity, increased profitability, and a sustainable competitive edge in today's dynamic industrial landscape.

CONCLUSION

TPM is an innovative approach that eliminates equipment breakdowns and defects. Integrating TPM with AI benefits industries by enhancing equipment maintenance and operational efficiency. This synergy shifts organizations from reactive to proactive maintenance, reducing unplanned downtimes through AI-driven predictive maintenance systems that monitor equipment health and preemptively address potential issues. Consequently, this minimizes unexpected stoppages, ensures uninterrupted production, and extends equipment lifespan, leading to lower capital expenditures and improved asset utilization.

The integration fosters a cultural shift toward continuous improvement and total employee involvement. AI empowers employees with insights, encouraging active participation in maintenance decisions, fostering a sense of ownership, and driving sustained improvements. The collaborative environment promoted by TPM, combined with AI's analytical capabilities, results in a motivated workforce committed to enhancing maintenance practices and overall performance.

Resource management and cost optimization are significantly improved through AI's precise planning and allocation of maintenance tasks. Appropriate planning reduces waste, lowers maintenance costs, and contributes to the organization's financial health and sustainability. Furthermore, AI-powered anomaly detection and real-time monitoring enhance workplace safety by reducing the risk of accidents.

In summary, integrating TPM and AI provides numerous advantages, including reduced downtimes, extended equipment lifespan, enhanced reliability, continuous improvement culture, better resource management, cost optimization, and improved operational safety. These benefits collectively lead to higher productivity, increased profitability, and a sustainable competitive edge in today's industrial landscape, positioning organizations for long-term success and resilience.

EXERCISES

1. What are the current system and process limitations of your TPM approach, and how can you apply AI to overcome these limitations?

TPM Phase	System and Process Limitations	AI Approach to Overcome System and Process Limitations
Phase 1: Inspecting target equipment		
Phase 2: Prioritizing identified problems		
Phase 3: Developing inspection standards and building PM schedules		
Phase 4: Training, implementing PM schedules, and monitoring		

2. What are the current cultural limitations of your TPM approach, and how can you apply AI to overcome these limitations?

TPM Phase	Cultural Limitations	AI Approach to Overcome Cultural Limitations
Phase 1: Inspecting target equipment		
Phase 2: Prioritizing identified problems		
Phase 3: Developing inspection standards and building PM schedules		
Phase 4: Training, implementing PM schedules, and monitoring		

3. What are some short- and long-term benefits to your department and organization for integrating AI into TPM?

#	Short-Term Benefits	Long-Term Benefits
Example	*Reduced unplanned downtimes*	*Sustained operational efficiency*
1		
2		
3		
4		
5		
6		

4. Reflect on the ethical implications of using AI in TPM within your organization. Consider aspects such as employee privacy, job displacement, and data security. Develop strategies to mitigate the negative aspects impact to the workforce.

#	Ethical Considerations	Strategies to Mitigate
Example	*Employee privacy concerns*	• *Ensure strict data privacy policies* • *Transparent data usage practices*
1		
2		
3		
4		
5		
6		

5. What are the technological considerations for integrating AI into your TPM framework within your organization? Develop strategies for the necessary technology integration.

#	Technology Necessary for AI	Strategy for Integration
Example	*Cloud computing resources*	• *Ensure scalable solutions* • *Secure cloud environments*
1		
2		
3		
4		
5		
6		

REFERENCES

Agustiady, T., & Cudney, E. (2023). *Total productive maintenance: Strategies and implementation guide* (2nd ed.). New York, NY: CRC Press.

Alatrista-Corrales, A., Moreno-Arévalo, M., Zevallos-Pacheco, C., & Rueda-Enríquez, M. (2021). Managing manufacturing efficiency using the concept of automation with human touch. *International Journal of Business Intelligence and Data Mining*, *19*(1), 33–51.

Alseiari, A., & Farrell, P. (2021). Optimising maintenance strategies through integrated artificial intelligence (AI) applications and total productive maintenance (TPM) for developing and enhancing the asset management of power networks. *International Journal of COMADEM*, *24*(1).

Antosz, K., Pasko, L., & Gola, A. (2020). The use of artificial intelligence methods to assess the effectiveness of lean maintenance concept implementation in manufacturing enterprises. *Applied Sciences*, *10*(21), 7922.

Barretiri, L. G. (2021). *Hospital operations management–a Lean healthcare approach to improve medical imaging management processes of a Portuguese public hospital* (Master's thesis, Universidade do Minho, Portugal).

Ćelić, J., Bronzin, T., Horvat, M., Jović, A., Stipić, A., Prole, B., Maričević, M., Pavlović, I., Pap, K., Mikota, M., & Jelača, N. (2024). Generative AI in E-maintenance: Myth or reality? *Artificial Intelligence*, *7*, 8.

Chen, X., & Voigt, T. (2020). Implementation of the manufacturing execution system in the food and beverage industry. *Journal of Food Engineering*, *278*, 109932.

Çınar, Z. M., Abdussalam Nuhu, A., Zeeshan, Q., Korhan, O., Asmael, M., & Safaei, B. (2020). Machine learning in predictive maintenance towards sustainable smart manufacturing in industry 4.0. *Sustainability*, *12*(19), 8211.

Cudney, E., Furterer, S., & Dietrich, D. (2013). *Lean systems*. New York, NY: CRC Press.

Cudney, E., Furterer, S., Laux, C., & Hundal, G. (2023). *Lean sustainability: A pathway to a circular economy*. New York, NY: CRC Press.

Fei, X., Bin, C., Jun, C., & Shunhua, H. (2020, August). Literature review: Framework of prognostic health management for airline predictive maintenance. In *2020 Chinese control and decision conference (CCDC)* (pp. 5112–5117). IEEE.

Kahraman, C., Çevik Onar, S., Öztayşi, B., & Karaşan, A. (2018). Operational planning in energy systems: A literature review. In Kahraman, C. & Kayakutlu, G. (Eds.), *Energy management – collective and computational intelligence with theory and applications* (vol. 149, pp. 335–356). Springer.

Korchagin, A., Deniskin, Y., Pocebneva, I., & Vasilyeva, O. (2022). Lean maintenance 4.0: Implementation for aviation industry. *Transportation Research Procedia*, *63*, 1521–1533.

Mohan, T. R., Roselyn, J. P., Uthra, R. A., Devaraj, D., & Umachandran, K. (2021). Intelligent machine learning based total productive maintenance approach for achieving zero downtime in industrial machinery. *Computers & Industrial Engineering*, *157*, 107267.

Pokvić, L. G., Deumić, A., Softić, A., & Badnjević, A. (2023). Application of artificial intelligence for management of maintenance and prediction of performances. In *Inspection of medical devices: For regulatory purposes* (pp. 115–138). Cham, Switzerland: Springer Nature.

Pundir, A., Maheshwari, P., & Prajapati, P. (2022, August). Machine learning based predictive maintenance model. In *Proceedings of the 2nd Indian international conference on industrial engineering and operations management* (pp. 2160–2174). https://ieomsociety.org/proceedings/2022india/528.pdf

Rabathi, S. R., & Hasbullah, H. (2023, August). Improving workflow of aircraft maintenance for reduce lead-time on nine-passenger aircraft. In *AIP conference proceedings* (Vol. 2485, No. 1). AIP Publishing.

Ruhiyat, Y. H., Sumaryo, S., & Susanto, E. (2022, December). Predictive maintenance for a ventilator using LSTM algorithm. In *2022 IEEE Asia Pacific conference on wireless and mobile (APWiMob)* (pp. 1–4). IEEE.

Siray, O. (2018, August). Prediction of industry 4.0's impact on total productive maintenance using a real manufacturing case. In *Proceedings of the international symposium for production research 2018* (p. 136). Springer.

Sukma, D. I., Setiawan, I., & Purba, H. H. (2021). A systematic literature review of overall equipment effectiveness implementation in Asia. *Journal of Industrial & Quality Engineering p-ISSN*, *2303*, 2715.

Tjernberg, L. B. (2023). Reliability-centered asset management with models for maintenance optimization and predictive maintenance: Including case studies for wind turbines. In *Women in power: Research and development advances in electric power systems* (pp. 87–155). Cham: Springer International Publishing.

Uzoigwe, D. O. (2024). Evaluating the effectiveness of reliability-centered maintenance programs in food and beverage manufacturing facilities: A review. *International Journal of Research and Innovation in Applied Science*, *IX*(II), 204–232.

Viharos, Z. J., Erdősi, G., Kovácsi, A., & Monostorii, L. (2010, May). Artificial intelligence applications for maintenance oriented health monitoring of wind turbines. In *Proceedings of the 7th conference on mechanical engineering* (pp. 403–413) https://citeseerx.ist.psu.edu/document?repid=rep1&type=pdf&doi=c7aed45fe1df1dda6d8761a8d6a10262849883c8.

10 Material Management-AI

FLOW, PULL, AND KANBAN OVERVIEW

Flow, pull, and kanban are foundational principles of Lean that create a system designed to optimize processes, reduce waste, and improve operational efficiency. These principles are essential for achieving a smooth and responsive process that can adapt to changing customer needs without unnecessary delays or excess inventory (Cudney et al., 2013).

Flow refers to the seamless progression of work through the process, ensuring that materials, products, and tasks move continuously from one stage to the next without interruptions, delays, or bottlenecks (Sanders et al., 2016). A well-maintained flow reduces waiting times and idle periods, allowing each component of the process to be completed as soon as the preceding task is finished. The ultimate goal of flow is to create an environment where work is constantly moving, which minimizes downtime and inefficiency. When processes flow smoothly, resources are utilized more effectively, lead times are shortened, and productivity improves.

In Lean, flow can be achieved through one-piece flow or batching. One-piece flow emphasizes the movement of individual tasks or units through each step of a process, where work is done on a single item or task at a time. This method minimizes delays between process stages and immediately exposes inefficiencies or quality issues, which enables quicker adjustments (Yavuz et al., 2006). One-piece flow reduces work-in-progress (WIP) or backlogs, leading to shorter cycle times, less waste, and a more responsive system.

In contrast, batching involves processing multiple units or tasks at once before moving them to the next stage of the process. While this approach may seem more efficient in specific contexts, especially in processes with high setup or preparation times, it often leads to higher levels of WIP, longer cycle times, and increased resource costs. Batching also increases the likelihood of delays since one batch must be completed before another can progress. Defects discovered within a batch can result in higher rework and correction costs compared to one-piece flow, where issues are identified sooner and isolated to a single unit or task.

By adopting a one-piece flow where feasible, organizations can improve the overall flow of tasks or services through the system, reduce waste, and increase operational efficiency. However, a hybrid approach may be necessary in some situations, particularly when setup times or process constraints make batching more practical.

In contrast to traditional push systems, where processes are based on forecasts or pre-set schedules, pull systems ensure that work is only triggered by actual customer demand. A pull system allows a task to be completed or resources to be allocated

DOI: 10.1201/9781003532972-10

only when there is an actual, concrete request from the customer. This approach reduces the risk of overproduction or over-allocation, a common form of waste in many industries. By providing only what is needed when it is required, organizations can maintain lower resource levels, which reduces storage or handling costs and minimizes the risk of obsolescence (Antony et al., 2022). The pull approach is particularly useful in environments with fluctuating demand because it allows operations to be more closely aligned with real-time customer needs.

Kanban, which originates from the Japanese word for "signboard" or "card," is a visual system that helps manage workflows by signaling when specific actions must be taken. In a kanban system, visual cues such as cards or bins notify team members when tasks must be completed or resources need replenishing. This system allows teams to track real-time progress, which ensures that tasks or resources are available exactly when needed without relying on complex scheduling (Barclay et al., 2022). Kanbans help balance work with demand. This balance enables organizations to avoid both shortages and excess resources and ensure that processes flow smoothly and efficiently through the system.

When the principles of flow, pull, and kanban are combined, they enable organizations to operate under the JIT methodology, where tasks or services are delivered exactly when needed and in the correct quantity. This approach minimizes waste by eliminating unnecessary work, reducing excess resources, and preventing assets from being tied up in unused inventory. Additionally, flow, pull, and kanban systems improve flexibility, which enables organizations to quickly adapt to changes in demand without sacrificing efficiency. By focusing on real-time needs and maintaining a continuous, demand-driven process, organizations can achieve higher levels of responsiveness and customer satisfaction while optimizing their use of resources and minimizing waste.

ADVANTAGES OF FLOW, PULL, AND KANBAN

Integrating flow, pull, and kanban systems offer several advantages that enhance operational efficiency and align services with customer demand. One of the most significant benefits is the reduction of waste. By closely aligning processes with actual demand, these systems help minimize common sources of waste, such as overproduction, excess resources, and unnecessary waiting times (Mojarro-Magaña et al., 2018). Tasks are triggered by actual customer needs rather than forecasts. Resources are used more efficiently. In addition, the risk of overallocation or unused assets is significantly reduced. These improvements lead to a more streamlined process with fewer wasted materials and efforts.

Another advantage is the increased flexibility these systems provide. Flow, pull, and kanban allow organizations to quickly adapt to changing customer demands. With pull systems and kanban in place, organizations can adjust schedules in real time to meet immediate needs. This flexibility is especially important in environments where customer preferences or service volumes fluctuate frequently because it allows organizations to stay responsive without accumulating excess resources or slowing down operations.

The improved responsiveness enabled by these systems ensures that actual customer demand, rather than forecasts, drives processes. These improvements mean organizations can respond more rapidly to market changes, whether ramping up services to meet increased demand or scaling back when needs decrease. By maintaining a continuous flow of work, organizations can adapt to new demands without disrupting operations or creating bottlenecks.

Enhanced visibility is another key advantage, particularly with kanban systems. Kanban provides real-time information on the workflows and resource requirements, which allows teams to monitor progress and inventory levels more accurately. This increased visibility improves communication across teams and departments, which makes it easier to coordinate efforts and avoid delays or shortages.

Finally, the cost savings associated with flow, pull, and kanban systems are significant. Organizations can save on storage, handling, and other operational costs by reducing excess inventory or resource allocation and avoiding overproduction. Additionally, fewer resources are wasted, which contributes to overall cost reductions. These systems help optimize efficiency and improve financial performance by eliminating unnecessary expenses and maximizing resource use.

LIMITATIONS OF FLOW, PULL, AND KANBAN

While flow, pull, and kanban systems offer significant benefits, they also have limitations that can impact their effectiveness. One key challenge is their manual dependency. Traditional pull and kanban systems rely on manual signals and monitoring to trigger actions such as task scheduling or resource replenishment. This reliance on manual processes can lead to delays, errors, or missed signals, which disrupt the timely flow of work and result in inefficiencies (Alaidaros et al., 2021). Human error or oversight can further exacerbate these issues, making maintaining a smooth and uninterrupted process challenging.

Another limitation is the lack of predictive capabilities in traditional systems. Flow, pull, and kanban are primarily reactive, meaning they respond to current conditions or customer demand but do not proactively anticipate future needs. Without predictive insights, organizations may struggle to prepare for potential disruptions or changes in demand, which can lead to missed opportunities or sudden bottlenecks in the workflow. This inability to foresee changes can cause delays and create inefficiencies.

The complexity in dynamic environments is another significant limitation. In industries or settings with frequent changes in customer demand or highly variable requirements, traditional flow and pull systems may find it challenging to keep pace. These systems are designed for more stable and predictable environments; therefore, they can become inefficient when demand fluctuates or when there are frequent changes in orders or processes. As a result, organizations may experience delays, increased cycle times, or mismatches between output and customer needs.

Visibility gaps also present a challenge for traditional systems. In environments where data is not continuously updated or monitored, visibility into resource levels and task status can be limited. Without real-time data, organizations may struggle to make informed decisions about when to schedule tasks or replenish resources, which can reduce the overall effectiveness of pull and kanban systems. This lack of transparency can lead to stockouts or overallocation, both of which negatively impact efficiency.

Finally, high resource costs can still be an issue with pull systems, particularly in high-volume operations. While pull systems aim to reduce overproduction, they may still require maintaining significant resource levels to meet customer demand. Managing this inventory or resource allocation can result in increased storage, handling, and financial costs, which could offset some of the efficiencies gained through flow, pull, and kanban systems.

While flow, pull, and kanban systems are effective tools for managing workflows, their manual dependency, reactive nature, and limitations in dynamic environments can hinder their full potential. Organizations must be aware of these challenges and work to mitigate them, possibly by integrating technology such as AI, which can enhance predictive capabilities, automation, and real-time visibility in these systems.

CULTURE

Employees may resist implementing flow, pull, and kanban systems for various reasons, many of which are rooted in organizational culture concerns and anxieties about changes to their work environment. A primary reason for resistance is fear of job displacement (Agustiady & Cudney, 2022). When these systems are introduced, particularly when integrated with automation or AI, employees may worry that their roles will be diminished or even replaced. These systems often streamline processes, reducing the need for manual labor in tasks such as inventory tracking or scheduling. Employees who have relied on or conducted these tasks as part of their daily responsibilities may fear their expertise is no longer valued, which can lead to uncertainty about their future in the organization.

Additionally, the shift from traditional, familiar processes to Lean-driven methods such as flow, pull, and kanban can feel disruptive, primarily in organizations that have operated under conventional push-based systems for a long time. Employees accustomed to working with pre-set schedules or forecasts may find it difficult to adjust to the JIT, demand-driven nature of pull systems. This transition requires employees to think and act more dynamically, responding to real-time signals rather than following a rigid schedule. This unpredictability can be stressful for some, particularly if they are used to having a clear and consistent workflow.

Another reason for resistance is that the increased visibility and accountability flow, pull, and kanban systems introduced. In traditional systems, inefficiencies or mistakes may go unnoticed for some time, especially in larger batch processes. However, in flow systems, problems become immediately visible because work moves continuously from one step to the next. Similarly, kanban boards and pull signals provide real-time data on inventory levels and workflow progress, which makes it easier for management to identify bottlenecks or underperformance. Employees may feel uncomfortable with this increased scrutiny, fearing that mistakes will be noticed more quickly. This concern can lead to anxiety about potential reprimands or a loss of job security.

Cultural resistance to change is another significant barrier when introducing these systems. In organizations with a more hierarchical, top-down management structure, principles such as employee empowerment and decentralized decision-making can be hard to embrace (Khan et al., 2024). Flow, pull, and kanban systems require frontline employees to take more responsibility for identifying problems and implementing solutions. If the organizational culture does not support this level of autonomy, or if employees are not used to actively participating in problem-solving, they may resist these changes. Employees may feel their input is not genuinely valued, which can diminish their sense of contribution. They may also lack confidence in their ability to contribute meaningfully to process improvements, which can lead to disengagement or passive resistance (Antony et al., 2023).

The lack of adequate training and support during the implementation of flow, pull, and kanban systems can also contribute to resistance (Barclay et al., 2022). These Lean tools often require a shift in mindset and skillset. Without proper guidance, employees may feel overwhelmed or unprepared to use them effectively if they are not given sufficient training to understand the purpose behind these systems and how to apply them. They may see the new processes as an added burden rather than an improvement, which can lead to frustration and reluctance to adopt the systems fully.

There can also be a misalignment between management and employee goals during the implementation of Lean practices. While management may prioritize efficiency and cost-saving benefits, employees may focus more on how these systems affect their daily tasks and job security. Suppose management fails to communicate how these changes will benefit the organization and the employees, such as by reducing unnecessary work or improving working conditions. In that case, employees may view the implementation as primarily serving the organization's interests at their expense.

Resistance to flow, pull, and kanban systems often stems from fear of job displacement, discomfort with increased accountability, resistance to change, and insufficient training. Addressing these concerns requires strong leadership, clear communication, and a commitment to building a supportive culture where employees feel empowered and valued as key contributors to the success of Lean initiatives.

INTEGRATING FLOW, PULL, KANBAN, AND AI

Integrating AI with traditional flow, pull, and kanban systems significantly enhances the efficiency and effectiveness of Lean processes. AI brings advanced automation, predictive capabilities, and real-time decision-making to these systems, allowing organizations to streamline operations and respond more effectively to changes in customer demand or process conditions. By leveraging AI, organizations can move beyond the limitations of manual processes and reactive systems, transforming their material management into a more dynamic, responsive, and data-driven operation.

One of the key areas where AI impacts Lean systems is flow optimization. In traditional Lean practices, maintaining smooth flow involves constant monitoring and manual adjustments to ensure that materials, products, and tasks move continuously through processes without bottlenecks or delays. With AI-driven tools, this process becomes automated and more precise. AI algorithms can analyze process flow in real time, detecting inefficiencies, identifying bottlenecks, and recommending immediate adjustments to keep operations running smoothly. For example, suppose a particular stage in the process starts to slow down. In that case, AI can identify the root cause, such as machine downtime or material shortages, and suggest interventions, such as rerouting tasks or rescheduling maintenance, to prevent further delays. This real-time analysis and optimization ensure that the flow is maintained, reducing downtime and improving overall process speed.

AI also improves pull systems by introducing predictive capabilities. Traditionally, pull systems rely on customer orders to trigger processes, which ensures no more products are made than needed. However, without the ability to anticipate future demand, traditional pull systems may struggle to keep pace with rapid market changes, which can lead to either stockouts or overproduction. AI-based demand forecasting models enhance the pull system by predicting customer demand based on historical data, market trends, and real-time factors such as seasonality or economic shifts. These AI models can anticipate fluctuations in demand, which enables organizations to adjust their process schedules proactively. This predictive ability reduces the risk of overproduction, which can lead to excess inventory. It also helps prevent underproduction, which can result in stockouts and unmet customer demand. By aligning processes more closely with predicted customer needs, AI-powered pull systems improve responsiveness to market changes while minimizing waste.

Integrating AI into kanban systems further automates and enhances material management through electronic kanban (e-kanban) systems. Traditional kanban relies on manual signals, such as cards or bins, to trigger replenishment orders when inventory levels reach a certain threshold. While effective, this system can experience delays or errors when signals are not processed in real time, leading to stock shortages or process slowdowns (Pekarcikova et al., 2020). AI-powered e-kanban systems automate this process by continuously monitoring inventory levels using sensors or real-time data inputs. When stock levels fall below a predetermined point, the AI system automatically generates replenishment orders, ensuring materials are restocked immediately. This automation eliminates the need for manual tracking, reduces human error, and ensures that processes always have the necessary materials on hand, improving the system's overall efficiency.

Moreover, AI-enhanced e-kanban systems provide better visibility and control over inventory levels. By integrating with enterprise resource planning (ERP) systems and supply chain management tools, AI can track internal inventory, supplier lead times, and delivery schedules. This comprehensive visibility allows organizations to optimize material availability and minimize disruptions caused by late deliveries or supply chain delays. The result is a more reliable, streamlined process where materials flow seamlessly through the system, reducing downtime and ensuring customer schedules are met on time.

Integrating AI into flow, pull, and kanban systems elevates traditional Lean practices by automating key processes, enhancing predictive capabilities, and enabling real-time decision-making. AI-driven flow optimization improves process efficiency by detecting and addressing bottlenecks as they occur. Predictive pull systems allow organizations to anticipate customer demand and adjust operational schedules accordingly, reducing waste and improving responsiveness. Further, AI-powered e-kanban systems automate material replenishment, ensuring that inventory levels are optimized and materials are available when needed. Together, these AI-enhanced systems create more dynamic, responsive, and efficient operations that align closely with Lean principles while leveraging the benefits of AI.

INTEGRATING FLOW, PULL, AND KANBAN WITHIN THE LEAN-AI MODEL

This integration allows organizations to streamline their processes, improve decision-making, and adapt to changing market conditions in a way that was not possible with traditional Lean systems alone. By leveraging AI's automation and predictive capabilities, organizations can achieve greater responsiveness, reduce waste, and optimize operations and material management practices.

At the tactical level, AI systems automate many of the routine tasks associated with flow, pull, and kanban, leading to faster response times and fewer manual errors. For example, in traditional Lean systems, inventory replenishment or operational scheduling often depends on manual inputs, which can introduce delays or mistakes. By automating these processes, AI ensures that tasks such as ordering materials or scheduling process runs are completed more quickly and accurately. Real-time data analytics further enhances this automation by providing constant visibility into material levels and operational status. In AI-driven kanban systems, for example, sensors and predictive algorithms monitor inventory levels continuously and automatically trigger replenishment orders when the stock falls below a certain threshold. These systems reduce the need for manual tracking and minimize the risk of stockouts or overstocking, keeping processes running smoothly. By automating these routine tasks, organizations can reduce the time and effort spent on manual monitoring and corrections, freeing up employees to focus on more value-added activities such as problem-solving and process improvement.

At the operational level, AI-powered tools provide deeper insights into operational trends and customer demand, enabling more informed decision-making and process optimization. In traditional Lean practices, decisions about operational schedules and inventory levels are often based on historical data or manual forecasts, which may not fully reflect current market conditions. However, with AI, organizations can analyze real-time data to predict shifts in customer demand and adjust their operational schedules accordingly. AI-driven demand forecasting tools, for example, can detect trends based on consumer behavior, seasonal patterns, or external factors such as economic conditions, allowing organizations to align their operations more closely with actual demand. These forecasting tools reduce the risk of overproduction, which can lead to excess inventory and waste, or underproduction, which can result in missed sales opportunities. By optimizing workflows, AI ensures organizations maintain a Lean and efficient operational system that responds dynamically to market conditions while minimizing waste.

At the strategic level, AI allows organizations to develop long-term material management and operations strategies by analyzing historical data and identifying patterns that may not be immediately apparent through manual analysis. Over time, AI systems gather vast amounts of data related to operational cycles, supply chain performance, and customer demand. By applying machine learning algorithms to this data, organizations can uncover trends and opportunities for improvement that would otherwise go unnoticed. For example, AI can identify recurring bottlenecks in the processes, areas where lead times could be shortened, or shifts in customer demand that suggest new product lines or adjustments to existing ones. These insights allow organizations to make more informed strategic decisions about allocating resources, optimizing workflows, and planning for future growth.

In addition, AI ensures that flow, pull, and kanban systems remain flexible and responsive to future market demands. In a rapidly changing global marketplace, where customer preferences and supply chain conditions can shift quickly, organizations must have systems in place that can adapt. AI-powered predictive models enable organizations to anticipate changes in the market and adjust their operations before problems arise. For example, suppose an AI system detects an impending increase in customer demand based on early indicators. In that case, it can recommend ramping up operations or adjusting inventory levels in advance, ensuring the organization is prepared to meet the demand without delays or stockouts. This proactive approach to material management and operational planning helps organizations stay competitive by maintaining the agility needed to respond to both short-term fluctuations and long-term trends.

Integrating flow, pull, and kanban systems with AI within the Lean-AI model provides significant benefits at the tactical, operational, and strategic levels, as shown in Figure 10.1. At the tactical level, AI automates routine tasks and enhances the accuracy of kanban systems, reducing manual errors and improving response times. At the operational level, AI-powered predictive tools provide deeper insights into operational trends and customer demand, enabling better decision-making and process optimization. Finally, at the strategic level, AI helps organizations develop long-term material management strategies by analyzing historical data and ensuring their systems remain flexible and responsive to future market demands. Integrating AI into Lean practices enhances efficiency and positions organizations to adapt to changing market conditions and drive long-term success.

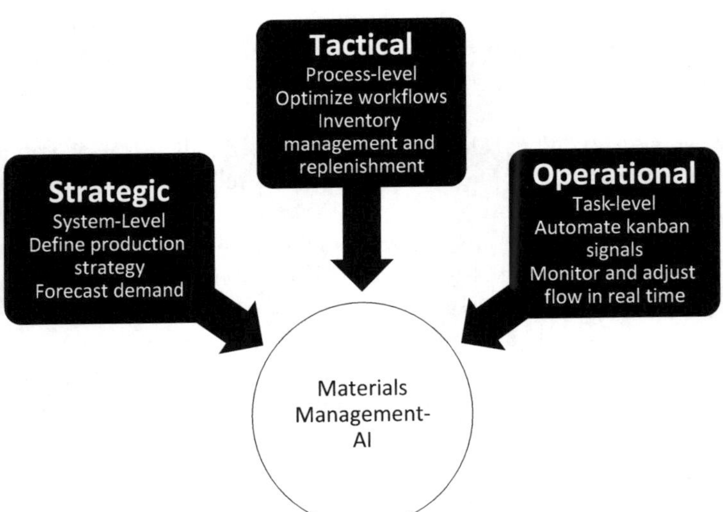

FIGURE 10.1 Materials management-AI model.

MATERIAL MANAGEMENT-AI DATA PRIVACY AND ETHICS CONSIDERATIONS

AI data privacy and ethics considerations in materials management revolve around safeguarding sensitive data, ensuring compliance with regulatory requirements, and maintaining ethical and fair business practices. These considerations span strategic, tactical, and operational levels, influencing how AI is implemented across material management processes.

At the strategic level, decisions shape the long-term direction of the organization, focusing on policies, governance frameworks, and compliance with privacy regulations. Organizations must establish clear data governance policies outlining the collection, storage, and processing of materials-related data while ensuring compliance with global privacy laws such as the General Data Protection Regulation (GDPR) and the California Consumer Privacy Act (CCPA). Additionally, defining data ownership is critical, particularly when AI models use information from third-party vendors or shared supplier networks. Bias mitigation is another key concern at this level, requiring organizations to develop strategies that identify and eliminate biases within AI algorithms. These biases can lead to unfair prioritization of certain suppliers, materials, or regions, ultimately distorting decision-making. Ensuring diversity in datasets helps create equitable supplier relationships and promotes fairness in procurement. Transparency and accountability must also be embedded into AI-driven materials management by requiring AI systems to clearly explain decisions, thereby fostering trust. Establishing an ethics committee to oversee AI applications in procurement and materials planning further reinforces accountability. Sustainability goals should also align with AI-driven materials management to optimize resource usage, reduce waste, and ensure ethical sourcing of materials, promoting responsible business practices.

At the tactical level, AI implementation focuses on bridging strategic goals with daily operations. Protecting supplier data is crucial, particularly during contract negotiations or performance evaluations, requiring robust encryption methods and secure data-sharing protocols to prevent unauthorized access. AI models used for vendor selection and risk assessments must be designed with fairness in mind, ensuring that supplier performance predictions and inventory optimization decisions do not disproportionately disadvantage certain suppliers or regions. Regular audits of AI algorithms help identify and correct any disparities in their recommendations. Data minimization is another essential aspect at this level, ensuring that AI models only use the necessary data to function effectively without excessive data collection. Anonymization techniques should be applied when integrating external supplier or logistics data to protect confidentiality. Furthermore, AI-driven risk management systems can proactively identify geopolitical risks, labor violations, or supplier non-compliance in the supply chain. However, organizations must ensure that these AI-powered risk assessments do not unfairly stigmatize suppliers or create biases that may exclude certain businesses without valid justification.

At the operational level, AI is integrated into daily material management activities, requiring strict data privacy protocols. Real-time data protection is essential for safeguarding information related to inventory levels, material usage, and supplier transactions. Implementing role-based access controls helps restrict sensitive data to authorized personnel, reducing the risk of data breaches. Ethical considerations also extend to AI-driven automation, particularly in tasks such as demand forecasting and reorder point calculations. While AI can improve efficiency, human oversight remains necessary to ensure ethical decision-making, especially in supplier selection and material substitution. Organizations must also prevent unethical practices, such as exploiting supplier vulnerabilities identified through AI analytics or using AI to bypass fair trade practices and labor rights. AI-powered workflow automation must be monitored to prevent favoritism or inefficiencies in task assignments, ensuring that resource allocation remains unbiased and equitable. Regular retraining of AI models with updated and diverse datasets helps prevent unintended biases from creeping into automated decision-making processes.

Across all levels, certain cross-functional considerations must be continuously addressed. Regulatory compliance must be actively monitored, with organizations adapting their AI-driven material management processes to evolving privacy and ethics laws. Employee training is another crucial component, ensuring that staff members at all levels understand ethical AI usage and best practices for data privacy. Implementing continuous audits of AI systems helps organizations assess compliance, detect biases, and refine ethical considerations over time. Lastly, collaboration with stakeholders – including suppliers, regulators, and internal teams – is essential to maintaining alignment on privacy and ethical standards in AI-driven materials management.

By embedding these considerations at strategic, tactical, and operational levels, organizations can harness AI in materials management responsibly and sustainably while fostering trust among stakeholders. Properly addressing AI-related privacy and ethical challenges not only ensures compliance but also reinforces the organization's commitment to fair, transparent, and socially responsible business practices.

FLOW, PULL, AND KANBAN-AI INTEGRATION INDUSTRY EXAMPLES

Integrating AI into various industries has transformed how flow, pull, and kanban systems operate, driving significant efficiency, responsiveness, and resource management improvements. These advancements are evident across multiple sectors, including automotive, electronics, pharmaceuticals, and retail, where AI has optimized processes and supply chains.

In the automotive industry, AI-powered kanban systems have enhanced inventory management on assembly lines. These systems automatically track part consumption in real time, allowing for precise and timely replenishment of materials (Marialuisa et al., 2023). By analyzing usage patterns on the production line, AI triggers replenishment orders exactly when needed, ensuring that the correct parts are available without causing delays (Mokudai et al., 2021). This approach has significantly reduced stockouts, which can halt production while also minimizing excess inventory. The result is a more efficient assembly process with lower inventory holding costs and improved production continuity.

In electronics manufacturing, AI-driven flow systems have optimized production by continuously analyzing production cycles and making real-time adjustments to machine scheduling (Lee et al., 2007). AI algorithms monitor equipment performance and predict potential downtime, allowing preemptive maintenance to be scheduled before a breakdown occurs (Tîțu & Bogorin-Predescu, 2022). This proactive approach reduces the likelihood of unplanned disruptions, keeping production on track and ensuring that machines always operate at peak efficiency. By predicting maintenance needs and optimizing scheduling, AI helps electronics manufacturers minimize downtime, increase productivity, and maintain smoother production flows.

AI has enhanced pull systems in the pharmaceutical industry by improving demand forecasting for various drugs. AI analyzes real-time patient needs, market trends, and historical data to predict fluctuations in drug demand more accurately (Tetteh-Caesar et al., 2024). AI-driven pull systems ensure that production schedules align closely with actual consumption, reducing the likelihood of overproduction or shortages. By anticipating demand, AI helps pharmaceutical organizations reduce waste from excess inventory while improving supply chain responsiveness to ensure patients receive medications when needed (Abideen & Mohamad, 2021), which is particularly valuable in the pharmaceutical sector, where timely delivery can be critical to patient health.

In the retail industry, AI-powered kanban systems are key in managing supply chains by predicting demand spikes and automatically adjusting replenishment schedules (Toe Teoh & Jin Goh, 2023). AI analyzes purchasing patterns, seasonal trends, and real-time sales data to forecast when demand will surge, particularly during peak sales. By adjusting replenishment schedules accordingly, AI helps retailers maintain optimal stock levels, reducing the risk of out-of-stock situations and ensuring that products are available when customers need them (Rathor, 2020). This capability improves stock availability and enhances customer satisfaction while minimizing the costs of carrying excess inventory.

FLOW, PULL, AND KANBAN-AI ADVANTAGES

Integrating AI into flow, pull, and kanban systems offers numerous advantages that significantly enhance the effectiveness and efficiency of material management in Lean operations. One of the most important benefits is real-time visibility. AI continuously monitors and analyzes inventory and operational data, providing real-time insights into material flow and stock levels. This level of visibility allows organizations to identify potential issues early and make immediate adjustments, reducing the risk of delays or disruptions in operations. By aligning processes more closely with actual demand, AI ensures that resources are utilized more effectively and that inventory levels are optimal, preventing shortages and excess.

Another key advantage is the improved efficiency that AI brings to replenishment processes. AI-driven kanban systems automate material replenishment, ensuring materials are available when needed. This automation eliminates manual tracking and reduces the risk of human error, which in turn minimizes downtime in processes and production lines. By ensuring that operations always have the necessary materials, AI helps increase productivity and maintain smooth operational flow, allowing organizations to meet demand more reliably.

Cost reduction is another significant benefit of integrating AI with flow, pull, and kanban systems. AI improves demand forecasting by analyzing historical data and current market trends, allowing for more accurate material needs predictions. With more precise demand forecasting, organizations can reduce excess inventory, leading to savings in storage and handling costs. This reduction in excess inventory also frees up capital that would otherwise be tied up in unused materials, improving overall financial performance.

AI also enhances the flexibility of flow, pull, and kanban systems, enabling organizations to adjust schedules and material flows quickly in response to changing customer demand. Adapting processes in real-time is crucial for maintaining competitiveness. AI-powered systems allow organizations to be more responsive and agile, making it easier to shift resources, modify production levels, and meet customer needs without sacrificing efficiency.

Finally, AI contributes to sustainability by optimizing resource use and reducing waste. AI-driven material management systems help organizations minimize overproduction, reduce excess inventory, and streamline processes, all of which reduce material waste. By improving efficiency and reducing waste, AI supports more sustainable practices, helping organizations meet environmental goals and reduce their environmental footprint. In this way, AI aligns Lean practices with operational efficiency and broader sustainability objectives, making material management more eco-friendly and responsible.

Integrating AI into flow, pull, and kanban systems offers real-time visibility, improved efficiency, cost reduction, enhanced flexibility, and contributes to sustainability. These advantages help organizations create more agile, efficient, and responsible processes that align with operational and environmental goals (Cudney et al., 2023).

CONCLUSION

Integrating AI into material management systems significantly enhances the application of Lean principles, particularly in flow, pull, and kanban systems. This convergence creates a powerful synergy, driving efficiency, responsiveness, and resource optimization improvements. By automating tasks such as inventory replenishment and process scheduling, AI reduces manual errors and frees up human resources for higher-value activities, ensuring steady process flow.

AI's predictive capabilities allow for more accurate demand forecasting, analyzing trends and market conditions to help organizations adjust operational levels preemptively. These predictive capabilities reduce overproduction and stockouts, ensuring leaner, more agile operations. AI also optimizes inventory levels by continuously monitoring stock, allowing organizations to maintain sufficient inventory without overcommitting resources to excess stock, further reducing waste.

A key advantage of AI is its ability to help organizations respond quickly to market changes. Real-time data analysis allows for dynamic decision-making, whether responding to shifts in demand or supply chain disruptions. This agility ensures that organizations meet customer expectations in a dynamic business environment.

Beyond operational gains, AI makes material management systems more adaptable to future challenges. AI-powered systems continuously learn and improve, ensuring long-term success as markets and technologies evolve. This adaptability allows organizations to remain resilient, aligning with Lean waste reduction and efficiency principles while positioning themselves for sustainable growth and competitiveness.

By integrating AI with Lean practices, organizations optimize current operations and build a future-ready system capable of adapting to the evolving business landscape, enabling long-term success.

EXERCISES

1. What are the current system and process limitations of your flow, pull, and kanban approach, and how can you apply AI to overcome these limitations?

Process Area	Current Limitation	AI Approach to Overcome System and Process Limitations
Example: Kanban replenishment	*Delays due to manual kanban signals*	*AI-powered e-kanban for automated, real-time replenishment*
Flow optimization		
Pull system demand response		
Inventory monitoring		
Process scheduling		

2. What are the current cultural limitations of your flow, pull, and kanban approach, and how can you apply AI to overcome these limitations?

Process Area	Current Limitation	AI Approach to Overcome Cultural Limitations
Example: Flow optimization	*Resistance to change due to fear of job displacement*	*AI training programs to show how AI complements human roles*
Kanban system visibility		
Pull system flexibility		
Real-time data integration		

3. Develop a high-level strategy for integrating AI into your flow, pull, or kanban system.

Process Area	AI Integration Strategy
Example: Kanban	*Implement AI-driven sensors to automate replenishment signals*

4. Reflect on the ethical implications of using AI in material management, such as privacy and job displacement. Propose solutions to mitigate these concerns.

Ethical Considerations	Impact	Strategies to Mitigate
Example: Job Displacement	*Employee morale*	*Upskill employees for AI-related tasks*

5. What are the technological considerations for integrating AI into your organization's flow, pull, and kanban framework? Develop strategies for the necessary technology integration.

#	Technology Necessary for AI	Strategy for Integration
Example	*Real-time data analytics for flow optimization*	*Build a robust data infrastructure*
1		
2		
3		
4		
5		
6		

6. Identify the tactical, operational, and strategic benefits of integrating AI into your material management processes.

Process Area	Tactical Benefits	Operational Benefits	Strategic Benefits
Example: Inventory Management	*Automated replenishment*	*Real-time inventory tracking*	*Long-term inventory optimization*

REFERENCES

Abideen, A., & Mohamad, F. B. (2021). Improving the performance of a Malaysian pharmaceutical warehouse supply chain by integrating value stream mapping and discrete event simulation. *Journal of Modelling in Management, 16*(1), 70–102.

Agustiady, T., & Cudney, E. (2022). *Building a sustainable Lean culture: An implementation guide*. New York, NY: CRC Press.

Alaidaros, H., Omar, M., & Romli, R. (2021). The state of the art of agile kanban method: Challenges and opportunities. *Independent Journal of Management & Production, 12*(8), 2535–2550.

Antony, J., Swarnakar, V., Cudney, E., & Pepper, M. (2022). A meta-analytic investigation of lean practices and their impact on organizational performance. *Total Quality Management & Business Excellence, 33*(15–16), 1799–1825.

Antony, J., Swarnakar, V., Gupta, N., Kaur, J., Jayaraman, R., Tortorella, G., & Cudney, E. (2023). Critical success factors for operational excellence initiatives in manufacturing: A meta-analysis. *Total Quality Management & Business Excellence, 34*(9–10), 1152–1172.

Barclay, R., Cudney, E., Shetty, S., & Antony, J. (2022). Determining critical success factors for lean implementation. *Total Quality Management & Business Excellence, 33*(7–8), 818–832.

Cudney, E., Furterer, S., & Dietrich, D. (2013). *Lean systems*. New York, NY: CRC Press.

Cudney, E., Furterer, S., Laux, C., & Hundal, G. (2023). *Lean sustainability: A pathway to a circular economy*. New York, NY: CRC Press.

Khan, L., Elshennawy, A., Cudney, E., & Furterer, S. (2024). Critical success factors for implementing industry 4.0 in aerospace and defense: A systematic literature review. *Quality Engineering*. https://doi.org/10.1080/08982112.2024.2403606

Lee, I. (2007). Evaluating artificial intelligence heuristics for a flexible Kanban system: Simultaneous Kanban controlling and scheduling. *International Journal of Production Research, 45*(13), 2859–2873.

Marialuisa, M., Matteo M. S., & Muhammad, S. (2023). Optimizing milk-run system and IT-based Kanban with artificial intelligence: An empirical study on multi-lines assembly shop floor. *Production & Manufacturing Research, 11*(1), 2179123.

Mojarro-Magaña, M., Olguín-Tiznado, J. E., García-Alcaraz, J. L., Camargo-Wilson, C., López-Barreras, J. A., & Pérez-López, R. J. (2018). Impact of the planning from the Kanban system on the company's operating benefits. *Sustainability, 10*(7), 2506.

Mokudai, T., Schröder, M., Müller, M., Schaede, C., Holst, H., Sinopoli, R., Jürgens, U., Herrigel, G., & Aoki, K. (2021). Digital technologies as lean augmentation: A preliminary study of Japanese automotive manufacturers. *International Journal of Automotive Technology and Management, 21*(3), 228–249.

Pekarcikova, M., Trebuna, P., Kliment, M., & Rosocha, L. (2020). Material flow optimization through e-kanban system simulation. *International Journal of Simulation Modelling, 19*(2), 243–254.

Rathor, K. (2020). Predictive modeling with AI for demand-driven distribution planning in ERP based on hybrid LSTM-DANN approach. *Innovative Computer Sciences Journal, 6*(1).

Sanders, A., Elangeswaran, C., & Wulfsberg, J. (2016). Industry 4.0 implies lean manufacturing: Research activities in industry 4.0 function as enablers for lean manufacturing. *Journal of Industrial Engineering and Management, 9*(3), 811–833.

Tetteh-Caesar, M. G., Gupta, S., Salonitis, K., & Jagtap, S. (2024). Implementing Lean 4.0: a review of case studies in pharmaceutical industry transformation. *Technological Sustainability, 3*(3), 354–372.

Tîțu, A. M., & Bogorin-Predescu, A. (2022). Contributions to the modeling of manufacturing processes for the implementation of the kanban methodology in the automotive industry. *International Journal of Mechatronics & Applied Mechanics*, (12), 94.

Toe Teoh, T., & Jin Goh, Y. (2023). AI in supply chain management. In *Artificial intelligence in business management* (pp. 225–255). Singapore: Springer Nature.

Yavuz, M., AKçali, E., & Tüfekçi, S. (2006). Optimizing production smoothing decisions via batch selection for mixed-model just-in-time manufacturing systems with arbitrary setup and processing times. *International Journal of Production Research, 44*(15), 3061–3081.

11 Hoshin Kanri-AI

HOSHIN KANRI OVERVIEW

Hoshin Kanri is a strategic decision-making tool that focuses resources on critical initiatives to accomplish organizational objectives. Hoshin Kanri is also commonly known as Policy Deployment. This process tool links major objectives with specific support plans throughout an organization (Cudney et al., 2013).

Policy deployment is a methodology to capture strategic goals and integrate these goals with your entire organization's daily activities. Figure 11.1 illustrates how policy deployment cascades down to all levels of the organization. The two levels of policy deployment include the following:

1. Management or strategic planning
2. Daily management

Senior leadership, including the chief executive officer (CEO) and directors, should use Hoshin Kanri to develop long-term strategic objectives. Mid-level managers should then use macro-level VSM to identify areas of improvement to achieve strategic goals. Finally, department teams should use Lean and Six Sigma tools for process improvement.

PLANNING FOR POLICY DEPLOYMENT

Effective planning is critical for the long-term success of a corporation. There are five main steps for effective planning:

- Step 1: Identify the organization's critical objectives
- Step 2: Evaluate the constraints
- Step 3: Establish performance measures
- Step 4: Develop an implementation plan
- Step 5: Conduct regular reviews

DAILY MANAGEMENT OF POLICY DEPLOYMENT

Daily management involves applying the plan-do-check-act (PDCA) to daily incremental continuous improvement to identify broad system problems in your organization. Once you gain a breakthrough improvement in the system problem, the improvement becomes the focus of daily continuous improvement activities. Hoshin

DOI: 10.1201/9781003532972-11

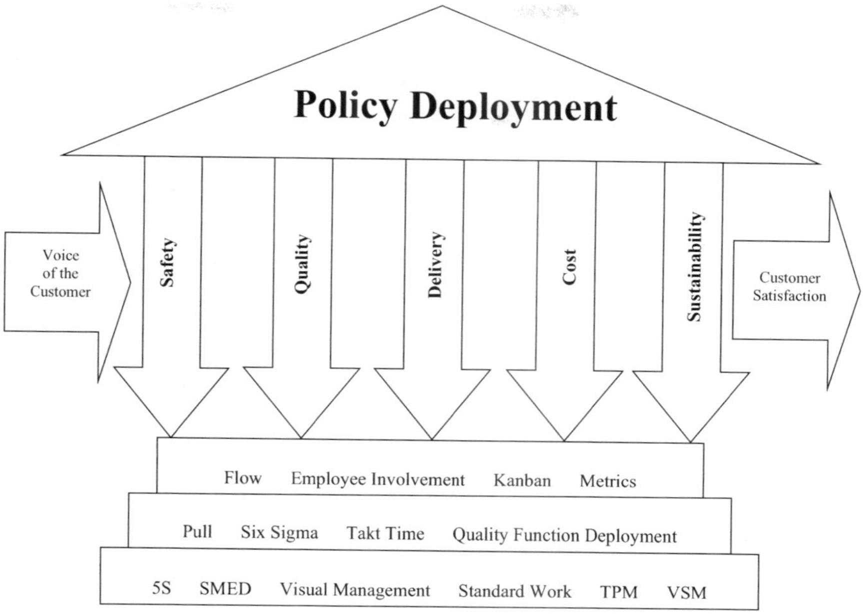

FIGURE 11.1 Strategic business system.

planning is the system that drives continuous improvement and breakthroughs. Policy deployment involves both planning and deployment (Cudney et al., 2023):

- Develop the targets
- Develop action plans to achieve the targets
- Deploy both

Maslow introduced the concept of a hierarchy of needs, outlining the basic needs that must be met before moving on to a higher need. Maslow's hierarchy of needs is illustrated in Figure 11.2. In concurrence with the hierarchy of needs that must be met for an organization to move on to its next higher need, there are five levels of organizational needs:

- Level 1: Core vision
- Level 2: Alignment
- Level 3: Self-diagnosis
- Level 4: Process management
- Level 5: Target focus

Figure 11.3 illustrates the five levels of organizational needs.

The five levels of organizational needs are directly linked to the six Hoshin planning steps and five Hoshin methods. Table 11.1 shows the linkage between the three.

Within the Hoshin planning process, there are six main steps.

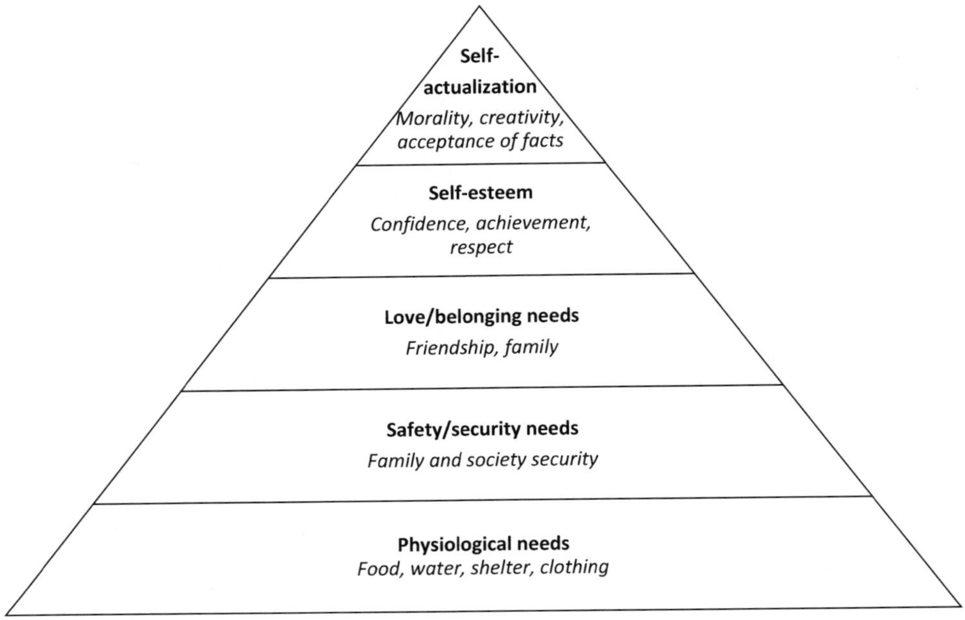

FIGURE 11.2 Maslow's hierarchy of needs.

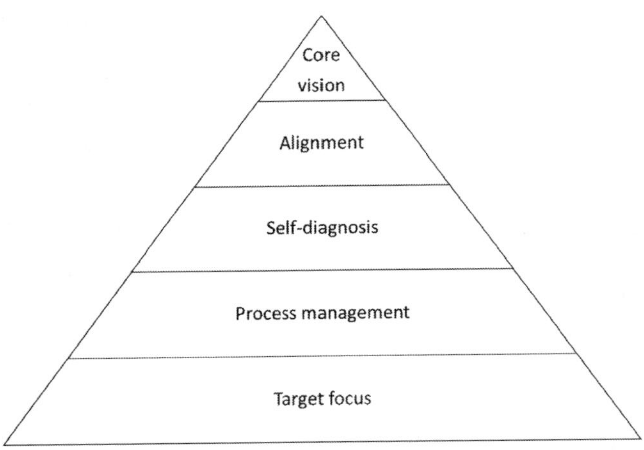

FIGURE 11.3 Levels of organizational needs.

STEP 1: DEVELOP A FIVE-YEAR VISION

Top management should develop a five-year vision to define the strategic objectives of your organization based on the internal, external, and environmental challenges your organization faces. The 5-year vision is the future target for your organization.

It is defined by your organization's goals, capabilities, and culture. It is a statement of where your organization wants to be.

TABLE 11.1
Linkage of Organizational Needs, Hoshin Planning Steps, and Hoshin Methods

Organizational Needs	Hoshin Planning Steps	Hoshin Methods
Core vision	5-year vision 1-year plan	Hoshin strategic plan summary
Alignment	Deployment	Hoshin plan summary
Self-diagnosis	Implementation	Hoshin action plan
Process management	Monthly reviews	Hoshin implementation plan
Target focus	Annual review	Hoshin implementation review

The vision is a communication tool for senior leadership to relate to the ideal future for the organization. For example, company XYZ has a vision to be a top 10 company in their field in 5 years.

STEP 2: DEVELOP A ONE-YEAR PLAN

Based on your five-year vision, you should develop a one-year plan to outline continuous improvement activities that will enable your company to achieve its long-term strategy. The one-year plan is linked to the five-year vision by taking an incremental step and defining key targets to attain that step. This step aims to focus the activities throughout all levels of your organization on addressing your external issues and improving your internal problems. As part of this step, you should analyze the external factors, including your competition and the economy as a whole. In addition, you should analyze past problems so your organization can avoid repeating them. Top management must prioritize the objectives based on safety, quality, delivery, and cost. Developing the one-year plan blazes the company's path to achieving its five-year vision.

STEP 3: DEPLOY THE ONE-YEAR PLAN

The next step is to deploy your one-year plan to all departments within your organization. Deploying your one-year plan is where you begin to set measurable goals for each department. The deployment step is a planning step to determine specific opportunities for improvement within each department. At this point, a strategy is set to achieve the path set by the one-year plan. This leads to step 4, which is implementation.

STEP 4: IMPLEMENT CONTINUOUS IMPROVEMENT ACTIVITIES

Each department must drive continuous improvement activities aligned with your one-year plan and five-year vision. This step is where the improvement process begins. The prior steps (1–3) involved planning for improvement; in step 4, you perform the improvement activities. For example, an improvement activity in this step could be an SMED event in a cell to reduce changeover time. The improvement activities in step 4 must tie directly back to the one-year plan. This step involves developing a master plan with appropriate measures and goals.

STEP 5: CONDUCT MONTHLY REVIEWS

You should track the progress of your continuous improvement activities using quantitative metrics and communicate them to senior leadership (CEO and directors) in a monthly review. The monthly reviews should link directly to your deployment of the one-year plan. You should monitor your actual improvements against your planned improvements as a monthly self-diagnosis and ensure the corrective actions are sustained. During a formal review to management, each department should present the tasks addressed, a problem analysis, and the problem-solving results. As the Hoshin planning implementation becomes more ingrained into your organization, the review may become more of a highlight or overview of the problems and corrective actions.

STEP 6: PERFORM AN ANNUAL REVIEW

Finally, you should conduct an annual review to monitor your progress and capture your organization's results. The annual review provides an opportunity to ensure that the implemented projects helped attain the one-year plan and the five-year vision. This checks how your implementation affected the organizational metrics set out in your five-year vision. Based on the previous year's results and the effectiveness of your implementations, a new one-year plan is developed to set targets and goals for the upcoming year. In addition, the organization may redevelop its five-year vision based on the current business environment.

ADVANTAGES OF HOSHIN KANRI

Implementing Hoshin Kanri offers several advantages to organizations striving for strategic alignment and operational excellence. One of the primary benefits is its ability to ensure that all levels of the organization are aligned with the company's long-term strategic goals. By integrating top-level strategic planning with day-to-day operations, Hoshin Kanri creates a clear and consistent direction for all employees. This alignment helps eliminate the disconnect between leadership's vision and the actions taken at various organizational levels, leading to more focused efforts and better resource utilization.

Another significant advantage of Hoshin Kanri is its emphasis on continuous improvement (kaizen) and innovation. The approach promotes regular reviews and adjustments to strategies based on actual performance data. This continuous feedback loop encourages organizations to be adaptable and responsive to changing market conditions, customer needs, and internal challenges. Hoshin Kanri helps companies stay competitive and continuously refine their processes and goals by fostering a culture of ongoing improvement.

Additionally, Hoshin Kanri empowers employees by involving them in the strategic planning process. Employees at all levels can contribute their insights and ideas through catchball (a collaborative communication process). This inclusive approach increases employee engagement and motivation and leverages the organization's collective knowledge, leading to more informed decision-making. Ultimately, this fosters a sense of ownership and accountability across the organization, which can significantly enhance execution and outcomes.

Finally, the structured and disciplined nature of Hoshin Kanri aids in reducing waste and inefficiencies. Organizations can allocate resources more effectively and ensure that their efforts drive meaningful results by focusing on the vital few objectives rather than being distracted by less critical tasks. This focus on critical success factors can enhance the organization's ability to achieve breakthrough goals and sustain long-term success.

LIMITATIONS OF HOSHIN KANRI

While Hoshin Kanri offers numerous advantages, it also comes with limitations that organizations must consider. One key challenge is the discipline and commitment required to implement the methodology effectively. Hoshin Kanri involves meticulous planning, execution, and ongoing review, which can be time-consuming and resource-intensive. Organizations that lack a strong culture of discipline and continuous improvement may struggle to maintain the rigorous focus and consistency needed for successful deployment. With this sustained effort, the benefits of Hoshin Kanri may be fully realized.

Another limitation is the complexity involved in coordinating efforts across different levels of the organization. Hoshin Kanri requires close collaboration and communication between top leadership and frontline employees to ensure alignment with strategic objectives. In large or siloed organizations, achieving this alignment can be difficult, as communication barriers and competing priorities may hinder the process. If alignment falters, the organization risks pursuing fragmented efforts that dilute the impact of the strategic plan. Additionally, cascading goals and ensuring buy-in from all levels can be challenging and time-consuming.

The long-term nature of Hoshin Kanri can also present challenges in rapidly changing environments. While the approach emphasizes strategic focus and alignment over extended periods, industries that experience frequent disruptions or require agility may find it challenging to maintain the relevance of their long-term goals. The need to periodically reassess and realign objectives can become cumbersome if market conditions shift dramatically and the organization may find itself locked into plans that no longer match its environment.

CULTURE

Cultural limitations are among the most significant challenges that organizations face when implementing Hoshin Kanri. This strategic planning method relies heavily on a culture of transparency, collaboration, and continuous improvement. If the organization's culture is not aligned with these principles, substantial obstacles to the successful implementation of Hoshin Kanri can be created (Agustiady & Cudney, 2022).

One of the main cultural limitations is resistance to change. Hoshin Kanri requires a shift in mindset from traditional hierarchical decision-making to a more inclusive, collaborative approach. Employees and leaders may resist this change in organizations with deeply ingrained top-down management structures. They may be accustomed to making decisions in isolation without soliciting input from others or may feel uncomfortable with the level of transparency and accountability required by Hoshin Kanri. Without a culture that embraces change and encourages participation, the implementation process can become stalled, and the intended benefits of Hoshin Kanri may not be realized.

Additionally, Hoshin Kanri emphasizes continuous improvement, which can be challenging in cultures where a "fixed mindset" prevails. In some organizations, there may be a belief that the current way of doing things is sufficient and that there is little need for ongoing reflection and refinement. This fixed mindset can hinder the adoption of Hoshin Kanri, as the methodology requires a commitment to regular review, feedback, and adjustment. Employees may view continuous improvement efforts as extra work or unnecessary if they do not see the value in consistently striving for better outcomes. This cultural resistance to continuous improvement can undermine the effectiveness of Hoshin Kanri.

Another cultural limitation is the organization's potential lack of empowerment and trust. Hoshin Kanri relies on the concept of "catchball," where ideas and feedback are thrown back and forth between different levels of the organization. This requires a culture of trust, where employees feel empowered to share their insights and are confident that their input will be valued. This collaborative process may break down in organizations with low trust, or employees may feel disconnected from leadership. Suppose employees do not believe that their voices matter. In that case, they are less likely to engage in the catchball process, which is essential for aligning strategies and ensuring the entire organization moves in the same direction.

Further, the emphasis on long-term strategic alignment in Hoshin Kanri can clash with cultures prioritizing short-term results. In some organizations, there is an intense focus on meeting immediate targets and achieving quick wins, often at the expense of long-term planning. This short-term orientation can conflict with the patient, methodical approach of Hoshin Kanri, which involves setting and pursuing long-term goals. Cultural resistance may arise when employees and leaders feel pressured to prioritize short-term performance over strategic alignment, leading to a lack of commitment to the Hoshin Kanri process.

Lastly, implementing Hoshin Kanri in multicultural or global organizations can present additional cultural challenges. Cultural backgrounds may influence how employees approach decision-making, collaboration, and feedback. For example, in some cultures, employees may be less comfortable openly discussing issues or challenging ideas from superiors, which can hinder the open communication and iterative feedback processes central to Hoshin Kanri. Navigating these cultural differences requires sensitivity and adaptability to ensure that Hoshin Kanri can be effectively implemented in diverse environments.

Overcoming these cultural limitations requires strong leadership and a deliberate effort to cultivate a culture that aligns with the principles of Hoshin Kanri. Leaders must be willing to invest in cultural transformation, fostering an environment of trust, empowerment, continuous improvement, and long-term thinking. Without this cultural foundation, organizations implementing Hoshin Kanri may face significant roadblocks and fail to deliver its full potential to drive strategic alignment and organizational success.

INTEGRATING HOSHIN KANRI AND AI

AI can be incorporated into the six main steps of Hoshin Kanri to enhance decision-making, streamline processes, and provide more accurate and actionable insights.

STEP 1: DEVELOP A FIVE-YEAR VISION

AI can significantly enhance the development of a 5-year vision by providing predictive analytics and scenario planning tools. By analyzing vast amounts of internal and external data, AI can identify emerging trends, potential risks, and opportunities that might impact the organization's future. This analysis can help top management define strategic objectives based on data-driven insights rather than relying solely on intuition or historical performance. AI can also simulate various scenarios, enabling leadership to assess the potential outcomes of different strategic directions and choose the most viable path forward. NLP tools can analyze external sources such as news, social media, and industry reports to gather insights into market trends, customer preferences, and competitor activities, further informing the five-year vision.

STEP 2: DEVELOP A ONE-YEAR PLAN

In developing a one-year plan, AI can analyze and forecast key performance indicators (KPIs) based on historical data and external factors, such as market conditions and economic trends. AI-driven tools can automate the analysis of large datasets to identify patterns and predict areas that require focus for continuous improvement. Machine learning algorithms help prioritize objectives by predicting which areas of the business will significantly impact achieving the five-year vision. AI can also perform risk assessments and recommend strategies that mitigate potential challenges, ensuring that the one-year plan is robust and adaptable to future uncertainties.

STEP 3: DEPLOY THE ONE-YEAR PLAN

During the deployment phase, AI can help set measurable goals for each department by analyzing performance data and identifying areas for improvement. AI-powered tools can be used to create and monitor detailed dashboards that track progress toward key targets. AI can also support decision-making at the departmental level by providing real-time insights and recommendations based on performance metrics. By leveraging AI to streamline data collection and analysis, organizations can ensure that their one-year plan is deployed effectively across all departments and performance is monitored and optimized.

STEP 4: IMPLEMENT CONTINUOUS IMPROVEMENT ACTIVITIES

AI can drive continuous improvement activities by automating routine tasks, analyzing real-time data, and providing predictive insights. For example, AI can optimize production processes by identifying inefficiencies and suggesting improvements, such as reducing waste, downtime, or cycle times. In manufacturing, AI can support initiatives like SMED by providing data-driven insights on process optimization. AI can also monitor the performance of continuous improvement initiatives and provide alerts or recommendations when deviations from expected outcomes occur, ensuring that improvement activities stay aligned with the one-year plan and five-year vision.

STEP 5: CONDUCT MONTHLY REVIEWS

AI can streamline monthly reviews by automating data collection and analysis, reducing the time required to prepare reports, and increasing the accuracy of performance tracking. Machine learning algorithms can identify trends and anomalies in the data that might not be immediately apparent through traditional analysis methods. AI can also generate visualizations and dashboards that provide real-time insights into how continuous improvement activities progress against the plan. Additionally, AI can help in root cause analysis by identifying underlying issues based on historical data and providing recommendations for corrective actions, which can be discussed during the monthly reviews with senior leadership.

STEP 6: PERFORM AN ANNUAL REVIEW

During the annual review, AI can help organizations comprehensively analyze their progress toward the one-year plan and the five-year vision. AI-powered tools can aggregate data from across the organization, analyze performance metrics, and identify areas where goals were met or missed. AI can also facilitate benchmarking by comparing the organization's performance to industry standards or competitors. Additionally, AI can support the development of the next one-year plan by providing predictive analytics that forecast future trends and suggest strategic adjustments. If necessary, AI can assist in redeveloping the five-year vision by simulating how different market conditions or internal changes might impact long-term objectives, ensuring that the organization remains adaptive and forward-looking.

Incorporating AI into the Hoshin Kanri process enables organizations to leverage data-driven insights for more effective planning, execution, and review. By integrating AI at each step, companies can improve decision-making, optimize resources, and stay agile in an ever-changing business environment.

INTEGRATING HOSHIN KANRI WITHIN THE LEAN-AI MODEL

Integrating AI into the Hoshin Kanri methodology combines tactical, operational, and strategic elements to enhance organizational alignment and drive continuous improvement across all levels, as shown in Figure 11.4. By leveraging AI, companies can optimize their strategic planning and execution processes, ensuring that short-term actions contribute meaningfully to long-term goals.

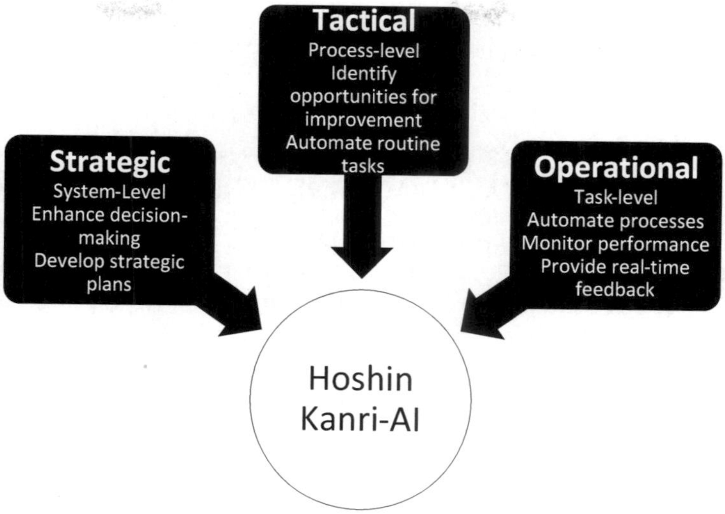

FIGURE 11.4 Hoshin Kanri-AI model.

On a tactical level, AI-driven data analytics play a crucial role in identifying opportunities for improvement. AI systems can analyze vast amounts of internal and external data to provide insights into potential risks, opportunities, and areas that require attention. This data-driven approach allows organizations to prioritize their improvement activities effectively and allocate resources to the areas with the most significant impact on achieving strategic goals. AI can also automate routine tasks, enabling teams to focus on high-priority initiatives that align with the company's long-term vision.

Operationally, AI facilitates the deployment and implementation of improvement activities by automating processes, monitoring performance, and providing real-time feedback. AI-powered tools can help departments track progress toward their measurable goals, ensuring they remain aligned with the overall strategy. By continuously analyzing data from operations, AI can identify bottlenecks or deviations from the plan, enabling teams to take corrective actions swiftly. This real-time monitoring and optimization ensure that the organization stays on course to achieve its annual and long-term objectives.

Strategically, AI enhances decision-making at the leadership level by providing predictive insights and scenario analysis. AI can help organizations develop more accurate and flexible strategic plans by simulating different scenarios based on market trends, competitor actions, and internal capabilities. This enables top management to refine their 5-year vision continuously, ensuring that it remains relevant in a rapidly changing environment. Furthermore, AI supports continuous review by analyzing performance data and offering actionable recommendations for future planning cycles, driving long-term success.

Integrating AI into Hoshin Kanri improves technical and operational processes and fosters a culture of collaboration and continuous improvement. By providing data-driven insights, AI empowers teams at all levels to contribute to strategic decision-making, creating a sense of ownership and involvement in the organization's success. This holistic approach combines tactical efficiency, operational precision, and strategic foresight, enabling organizations to achieve their long-term vision while continuously adapting to new challenges and opportunities.

HOSHIN KANRI-AI DATA PRIVACY AND ETHICS CONSIDERATIONS

Integrating AI into Hoshin Kanri brings numerous strategic advantages, but it also raises critical data privacy and ethical considerations that organizations must address. These considerations span across strategic, tactical, and operational levels, influencing decision-making, data governance, and ethical AI implementation within the policy deployment framework.

At the strategic level, AI-driven Hoshin Kanri relies on vast amounts of internal and external data to forecast trends, set long-term objectives, and analyze market conditions. Organizations must establish robust data governance policies to ensure compliance with global privacy laws, such as the GDPR and the California Consumer Privacy Act (CCPA). Clear guidelines must define the collection, storage, and usage of AI-driven insights to prevent unauthorized access or misuse of sensitive organizational data. Additionally, transparency and accountability are essential to prevent bias in AI-driven strategic planning. AI systems must be designed to explain their decision-making processes clearly, ensuring that leadership understands the basis of AI-generated recommendations. Ethical concerns also extend to bias mitigation, as AI algorithms used in scenario planning must be trained on diverse datasets to avoid reinforcing existing disparities. By embedding ethical AI principles into the strategic planning phase, organizations can ensure that AI-enhanced Hoshin Kanri aligns with fairness, compliance, and corporate responsibility.

At the tactical level, AI is used to support the deployment of the 1-year plan, allocate resources, and optimize continuous improvement activities. However, AI-driven automation in decision-making must be balanced with human oversight to avoid unintended consequences. One of the key concerns is algorithmic fairness, particularly in AI models that evaluate project priorities, allocate funding, or identify high-impact improvement initiatives. Regular audits should be conducted to ensure that AI does not disproportionately prioritize certain business units or teams over others. Additionally, data minimization is a crucial principle in AI-driven Hoshin Kanri, ensuring that only the necessary data is collected and analyzed to maintain compliance with privacy regulations. Organizations should also implement secure data-sharing practices when integrating AI insights into policy deployment across departments, protecting sensitive business strategies from cyber threats or leaks. By adopting AI in a way that enhances rather than replaces human decision-making, organizations can strike a balance between automation and ethical responsibility in tactical planning.

At the operational level, AI assists in real-time tracking of key performance indicators (KPIs), monitoring continuous improvement activities, and providing predictive insights for monthly and annual reviews. However, real-time data privacy is a significant concern when AI systems continuously collect and process employee performance metrics, workflow efficiencies, and organizational progress. Organizations must establish role-based access controls to ensure that only authorized personnel have access to AI-generated insights, preventing misuse or surveillance concerns. Additionally, AI automation ethics must be considered when AI recommends operational adjustments, such as reallocating resources or prioritizing improvement initiatives. AI should function as a decision-support tool rather than an autonomous decision-maker, preserving human judgment in policy deployment. Furthermore, ethical data usage must be maintained to prevent AI-driven insights from being misused to justify workforce reductions or performance-based disciplinary actions without proper human evaluation. Lastly, bias in workflow automation is a growing concern, as AI-powered project tracking and resource allocation tools must be regularly audited to prevent unintended favoritism or systemic inefficiencies. Continuous monitoring and retraining of AI models with updated, unbiased datasets can help organizations uphold ethical standards in AI-driven Hoshin Kanri.

Beyond these levels, certain cross-functional considerations must be continuously addressed. Regulatory compliance should be actively monitored to ensure that AI-driven Hoshin Kanri remains aligned with evolving privacy laws and ethical guidelines. Employee training is critical to educating teams on the responsible use of AI-driven policy deployment, fostering trust and transparency in AI-assisted decision-making. Continuous audits of AI models must be conducted to assess compliance, bias, and ethical considerations, ensuring that AI enhances rather than undermines strategic planning. Additionally, stakeholder collaboration is essential for aligning AI-driven Hoshin Kanri with organizational values, involving executives, employees, and regulatory bodies in discussions about AI's role in strategic decision-making.

By embedding these AI data privacy and ethics considerations into Hoshin Kanri, organizations can ensure that AI supports strategic alignment, enhances decision-making, and drives continuous improvement responsibly. A well-governed Hoshin Kanri-AI integration fosters trust among stakeholders while maintaining compliance, ethical fairness, and transparency in long-term organizational planning.

HOSHIN KANRI-AI INTEGRATION INDUSTRY EXAMPLES

Integrating Hoshin Kanri and AI offers transformative benefits across various industries by aligning strategic goals with operational excellence through data-driven insights. In manufacturing, companies utilize AI to analyze production data and optimize continuous improvement activities, ensuring alignment with their five-year vision of increased efficiency and reduced waste. In healthcare, organizations apply AI-powered analytics to patient care processes, enabling real-time adjustments that enhance service quality and operational performance while aligning with strategic objectives for patient outcomes. Retail companies leverage AI to track customer behaviors and market trends, feeding insights into their Hoshin Kanri processes to develop more accurate one-year plans and adjust their real-time strategies. Across industries, integrating Hoshin Kanri and AI drives operational improvements, enhances productivity, and ensures that organizations remain agile and responsive to changing market conditions, ultimately contributing to long-term success.

HOSHIN KANRI-AI ADVANTAGES

Integrating AI into the Hoshin Kanri process offers several significant advantages that enhance strategic planning and execution. One of the primary benefits is the ability to harness predictive analytics for more informed decision-making. AI can analyze vast amounts of internal and external data to identify trends, forecast future scenarios, and detect potential risks. This allows organizations to make data-driven decisions when developing their long-term vision and strategic goals, reducing reliance on intuition and historical data alone. AI can provide leadership with real-time insights, enabling more accurate predictions about market shifts, customer preferences, and competitive dynamics, which can be invaluable for setting a robust 5-year vision.

AI also streamlines the implementation and monitoring of strategic plans by automating routine tasks and improving efficiency. During the deployment of the 1-year plan, AI can automate data collection, performance tracking, and reporting processes, freeing up valuable time and resources. AI-powered dashboards can offer real-time updates on KPIs and continuous improvement activities, allowing organizations to monitor progress more effectively. This level of automation ensures that strategic goals are consistently aligned with day-to-day operations and that any deviations or emerging issues are promptly identified and addressed.

Further, AI enhances flexibility and responsiveness within the Hoshin Kanri framework. By leveraging machine learning algorithms, organizations can continuously analyze performance data and adjust their strategies based on real-time feedback. This analysis enables a more dynamic approach to strategy execution, where adjustments can be made quickly in response to changing conditions. AI can also aid in scenario planning, allowing organizations to simulate different strategic outcomes and test the impact of various decisions before implementing them. This adaptability ensures that the organization remains agile and resilient, even in volatile or uncertain environments.

Finally, AI supports a culture of continuous improvement by providing predictive insights and recommendations for process optimization. AI can analyze complex data sets to identify inefficiencies and suggest improvements, making it easier to drive process optimization efforts across the organization. Whether it's reducing waste in production, improving cycle times, or enhancing resource allocation, AI can deliver actionable insights that lead to more effective and efficient operations. This continuous flow of data-driven insights helps organizations achieve their strategic objectives and maintain a competitive edge by fostering innovation and improvement at all levels.

CONCLUSION

Hoshin Kanri, or policy deployment, can be significantly enhanced by integrating AI. This strategic methodology aligns an organization's major objectives with specific support plans at all levels. AI-driven tools can optimize each phase of the Hoshin Kanri process, from developing a long-term vision to implementing continuous improvement activities and conducting regular reviews. By incorporating AI, organizations can improve decision-making, increase agility, and ensure that strategic objectives align with real-time insights from vast data sources.

One significant advantage of AI in Hoshin Kanri is its capacity to deliver predictive analytics, which supports more informed decision-making when setting long-term goals. AI can analyze large volumes of internal and external data to identify trends, risks, and opportunities that might not be apparent through traditional methods. This data-driven approach allows leadership to craft more accurate strategic plans, ensuring that the organization's long-term vision remains relevant amidst changing market conditions.

AI also streamlines the deployment and execution of strategic plans. Automation tools can take over routine tasks like data collection and performance tracking, allowing for more efficient execution of the 1-year plan. AI-powered dashboards provide real-time updates on key performance indicators, ensuring that continuous improvement efforts are closely monitored and optimized. This level of automation allows for the timely identification of deviations and issues, ensuring that strategic objectives remain on track.

Moreover, AI enhances flexibility and responsiveness in Hoshin Kanri by enabling continuous analysis of performance data and real-time strategic adjustments. AI facilitates scenario planning, allowing organizations to test the impact of various decisions before implementation. This adaptability ensures organizations can quickly respond to evolving conditions, maintaining agility and resilience in unpredictable environments.

Finally, AI fosters a culture of continuous improvement by providing predictive insights that drive process optimization. AI can identify inefficiencies and suggest organizational improvements, supporting efforts to reduce waste, improve cycle times, and enhance resource allocation. This data-driven approach to improvement ensures that strategic objectives are consistently achieved while fostering innovation and keeping organizations competitive and forward-looking.

EXERCISES

1. What are the current system and process limitations in your Hoshin Kanri process, and how can you apply AI to overcome these limitations?

Hoshin Kanri Step	System and Process Limitations	AI Approach to Overcome System and Process Limitations
Step 1: Develop a 5-Year Vision		
Step 2: Develop a 1-Year Plan		
Step 3: Deploy the 1-Year Plan		
Step 4: Implement Continuous Improvement Activities		
Step 5: Conduct Monthly Reviews		
Step 6: Perform an Annual Review		

2. What are the current cultural limitations in your Hoshin Kanri process, and how can you apply AI to overcome these limitations?

Hoshin Kanri Step	Cultural Limitations	AI Approach to Overcome Cultural Limitations
Step 1: Develop a 5-Year Vision		
Step 2: Develop a 1-Year Plan		
Step 3: Deploy the 1-Year Plan		
Step 4: Implement Continuous Improvement Activities		
Step 5: Conduct Monthly Reviews		
Step 6: Perform an Annual Review		

3. What are some short- and long-term benefits for your organization when integrating AI into Hoshin Kanri?

#	Short-Term Benefits	Long-Term Benefits
Example	*Faster analysis of internal and external data for decision-making*	*Sustained alignment of strategic goals with market trends*
1		
2		
3		
4		
5		
6		

4. Reflect on the ethical implications of using AI within your Hoshin Kanri process. Consider aspects such as employee privacy, job displacement, and data security. Develop strategies to mitigate the negative impacts to the workforce.

#	Ethical Considerations	Strategies to Mitigate
Example	*Bias in AI-driven decision-making process*	*Implement AI audits to check for bias and ensure fairness*
1		
2		
3		
4		
5		
6		

5. What are the technology considerations for integrating AI into your Hoshin Kanri framework within your organization? Develop strategies for the necessary technology integration.

#	Technology Necessary for AI	Strategy for Integration
Example	*Cybersecurity measures to protect AI systems*	• *Develop AI-specific security protocols* • *Invest in AI threat detection tools*
1		
2		
3		
4		
5		
6		

REFERENCES

Agustiady, T., & Cudney, E. (2022). *Building a sustainable Lean culture: An implementation guide*. New York: NY: CRC Press.
Cudney, E., Furterer, S., & Dietrich, D. (2013). *Lean systems*. New York, NY: CRC Press.
Cudney, E., Furterer, S., Laux, C., & Hundal, G. (2023). *Lean sustainability: A pathway to a circular economy*. New York, NY: CRC Press.

12 Lean Office-AI

LEAN OFFICE OVERVIEW

As Lean manufacturing principles gained global attention, organizations began to explore their application in office settings, particularly in administrative and service processes. The focus was on improving workflow, reducing bottlenecks, and enhancing efficiency in areas like work order processing, document control, human resources, finance, and others.

The concept of "Lean office" began to formalize, as businesses began applying Lean tools such as 5S (sort, set in order, shine, standardize, sustain), poka-yoke, VSM, and kaizen (continuous improvement) to office processes. This led to the development of Lean administration, which focused on reducing lead times, errors, and waste in paperwork, decision-making, and information flow.

Thus, a Lean office became formalized as a set of principles, techniques, and philosophies used to improve the efficiency of administrative processes in an office (Papadopoulos, 2010). The goal of a Lean office is to eliminate waste and wasteful activities to increase quality, customer satisfaction, and efficiency in business processes.

Applying Lean principles to an office can help to improve processing time and reduce handoffs through the reduction or elimination of steps, activities, and processes. Business processes, similar to manufacturing processes, hide unnecessary tasks or touch points. Elimination of these unnecessary steps leads to greater control of the information flow within the office.

With the exponential spread of digitization and automation, Lean principles in the office evolved to include software and technology solutions. Organizations began using Lean to streamline digital workflows, improve data accuracy, and enhance collaboration across virtual teams. Concepts like Lean IT and Lean project management emerged during this time.

One of the most prolific adopters of Lean technology was in software development. Mary and Tom Poppendieck leveraged Lean manufacturing in their book, *Lean Software Development: An Agile Toolkit* (Poppendieck, & Poppendieck, 2003), which outlined how development teams could apply the methods to developing software. For example, bottlenecks in Lean software development may not result from machine downtime like in manufacturing but can result from inefficient handoffs between the development team and QA teams, The Lean practices help identify this bottleneck and allow us to adjust workloads and team capacity to eliminate the bottlenecks.

Advances in technology and changing workplace dynamics have driven Lean in the office and agile methodologies. The core principles of waste reduction, value creation, and continuous improvement remain central to its application in various office settings and software development.

DOI: 10.1201/9781003532972-12

Today, "Lean office" is an integral part of the operational strategies of most organizations that have adopted a Lean strategy. It is used to enhance efficiency and foster a culture of continuous improvement, innovation, and employee engagement. Lean office principles are now applied to remote work environments, knowledge work, and complex service industries, demonstrating their versatility and enduring relevance (Tapping & Dunn, 2006).

THE MAJOR DIFFERENCE IN TOOLS FOR LEAN IN MANUFACTURING AND LEAN IN OFFICE

See Table 12.1.

TABLE 12.1
The Major Difference in the Seven Wastes for Lean in Manufacturing versus Lean in Office

Lean in Manufacturing	Lean in Office
7 Wastes	**7 Wastes**
• **Transport**	• **Travel**
• **Inventory**	• **Inventory (backlog)**
• **Motion**	• **Motion**
• **Waiting**	• **Waiting**
• **Overproduction**	• **Over-documentation**
• **Overprocessing**	• **Overprocessing**
• **Defects**	• **Defects/Rework**
Value Stream Mapping	**Swim Lane Mapping**

AGILE SOFTWARE DEVELOPMENT

Agile was introduced in the late 1990s as software development needed a more efficient approach than a traditional waterfall approach to project management (Hughes & Cotterell, 2009). A waterfall approach is a single-pass software development methodology where requirements are defined at the start of the development process and tasks are completed in sequential phases of initiation, planning, execution, monitoring, control, and close-out. This rigid workflow became a roadblock to innovation as the requirements of software development shifted toward an ultra-rapid speed.

In 2001, a group of Agile practitioners came together for a summit in Utah and introduced the Agile Manifesto (Beck et al., 2001). The agile manifesto promised quicker release cycles, happier teams, and more satisfied customers. Although this methodology is still evolving, the Lean of Lean technology on agile is irrefutable.

LIMITATIONS OF LEAN OFFICE

Difficulty in Measuring Office Processes: In manufacturing processes, the outputs and waste are often visible, whereas office processes involve hidden, information and knowledge tasks that can be harder to visualize, measure, and standardize. This makes identifying inefficiencies and implementing Lean improvements more challenging in the Lean office.

Excessive Focus on Efficiency: Lean's focus on efficiency improvement and waste reduction can sometimes lead to an overemphasis on speed and productivity at the expense of creativity and innovation. In office environments, where problem-solving and strategic thinking are crucial, this can be a significant drawback.

Resistance to Change: Employees in office settings may resist Lean initiatives, particularly if they perceive them as a threat to their autonomy or job security. This resistance can hinder the successful adoption of Lean practices.

Misalignment with Information Work: Lean's roots in manufacturing make it well-suited for repetitive tasks, but office work often involves information-intensive activities that require flexibility and creativity. Applying Lean too rigidly in these contexts can stifle innovation and reduce job satisfaction.

Value Streams Mapping: In an office setting, value streams are often more complex and less linear than in manufacturing. This complexity can make it difficult to apply Lean tools like VSM effectively. Thus, Lean office applications require the use of a swim lane mapping to document the complexities and hand-offs.

Cultural and Organizational Barriers: For Lean office implementations to succeed, it is important to cultivate a culture of collaboration. Fostering this culture can be challenging in an office environment that has siloed departments and hierarchical structures.

Quick Wins: Lean office initiatives can sometimes focus on quick wins. However, Lean office results need a longer-term focus on sustained continuous improvement as sometimes transformations in office applications take longer to materialize. This can lead to discontent among leaders and result in a premature termination of the Lean office initiatives.

Not Applicable to some Creative and Highly Variable Processes: Lean office initiatives may not apply to some creative or highly variable processes where standardization and waste reduction are less relevant. In such cases, applying Lean office principles may not be suitable at all.

Dependent on Leadership Commitment: Due to delayed results, Lean office implementations require a strong and sustained commitment from leaders. Without this commitment, Lean office initiatives can lose momentum, leading to partial implementation or failure to achieve desired results.

Oversimplification: In the need to streamline and eliminate waste, there is a risk of oversimplifying processes, which can lead to the loss of valuable nuances and suboptimal decision-making.

These limitations indicate that while Lean technology is beneficial in office environments, its implementation requires careful consideration, adaptation to context, and a focus on balancing efficiency with creativity and employee engagement (Bhamu & Sangwan, 2014).

LIMITATIONS OF AGILE

Agile methodologies, while highly effective in many scenarios, come with some limitations, as seen below:

Customer Involvement: Agile emphasizes constant customer feedback and involvement. If customers are not available or willing to engage regularly, it can impede progress and affect the quality of the final product.

Scope Creep: Agile's flexibility can sometimes lead to scope creep, where continuous changes and additions to project requirements can lead to unmanageable project scope, delays, and budget overruns (Boehm, 2006).

Resource Intensity: Agile requires frequent collaboration, regular meetings (such as daily stand-ups), and active participation from all team members. This can be resource-intensive and may not always be feasible for all organizations or projects.

Requires Experienced Team Members: Agile methodologies often demand a high level of skill and experience from team members. Teams need to be self-organizing and capable of making decisions independently, which might be challenging for less experienced teams.

Documentation Challenges: Agile prioritizes working software over comprehensive documentation. While this can accelerate development, it can lead to insufficient documentation, which may pose challenges for future maintenance or new team members joining the project.

Inconsistent Results: The iterative nature of Agile can sometimes lead to inconsistent results if the feedback loops are not managed effectively. Projects might face challenges in meeting initial expectations if changes are not properly integrated.

Difficulty in Predicting Timelines: Agile's focus on iterative development and evolving requirements can make it challenging to predict project timelines and delivery dates with precision.

Overemphasis on Flexibility: The focus on adaptability and change can sometimes lead to a lack of clear direction and prioritization, which may result in inefficiencies or misalignment with the project's overall goals.

Addressing these limitations often involves balancing Agile principles with effective planning, communication, and strategic management practices.

CULTURE

Since Lean in office is more human-centric then Lean in manufacturing process, the focus on the two humanistic pillars of Lean – respect and humility are critical to the process of planning, execution, and sustainment of the Lean in office initiatives.

Due to the limitations of Lean office methodology, the culture perspective must be considered before every initiative. These limitations often stem from siloed departments, resistance to change, entrenched work practices, and a negative attitude toward process improvement (Netland, & Ferdows, 2014).

Resistance to Lean office implementations can be caused by several factors, including:

Lack of training: Inadequate training is one of the major challenges faced during Lean office implementations. Leaders may be reluctant to invest in training, resulting in employees not fully understanding the principles and practices of Lean office, leading to resistance and inefficiency.

Fear of change: When implementing a Lean office, employees may resist the initiative due to the fear of job insecurity, increased workload, or the need to adapt to new processes.

Lack of understanding: Some people may not fully understand the benefits of Lean in office applications or how it applies to their roles.

Comfort with the status quo: This resistance to change is often felt because employees believe that their existing processes are sufficient; therefore, they do not see the need for change.

Significant changes: Lean often requires significant changes to workflow and processes, which may lead to reorganizations and job rotations.

Management Resistance (Agile): Agile requires a shift in management and organizational culture. Traditional management practices may resist Agile methodologies, leading to challenges in implementation and adoption (Leffingwell, 2011).

INTEGRATING LEAN IN OFFICE AND AI

Integrating Lean principles in office settings with AI can transform operations, drive efficiency, and foster a culture of continuous improvement. This integration can have a major impact on people, processes, and organizational culture.

The impact on people includes empowerment, skill development, and engagement. Lean office encourages employee involvement in problem-solving and root-cause analysis. AI enhances this by providing tools that make data-driven decision-making possible, empowering teams to rapidly, and accurately identify inefficiencies and drive sustainable improvements.

Skill development and upskilling can be easier as repetitive tasks can be delegated to AI. This frees up employees to focus on creative, higher-value activities, necessitating upskilling. Learn-Office and AI transition nurtures a workforce that is more focused, innovative, cross-trained, and flexible.

Workforce engagement increases as automating repetitive and mundane tasks using AI allows employees to engage more deeply in their work, resulting in increased job satisfaction and reducing employee burnout.

The impact of integrating Lean office with AI on processes includes streamlining the operations, enhanced decision-making, continuous improvement, data-driven culture, streamlined processes, enhancing speed and accuracy, collaboration, transparency, and agility.

Application of Lean office-AI principles can streamline operations by eliminating waste (muda) in processes, and AI can automate and optimize these streamlined processes, enhancing speed and accuracy. Enhanced decision-making results from using AI for real-time insights and predictive analytics, enabling quicker and more informed decisions that align with the emphasis of Lean office-AI on data-driven management.

AI tools provide real-time, continuous process monitoring and identification of areas for improvement, ensuring that Lean office methodologies are consistently applied making continuous improvement an ongoing initiative. Integration of AI with Lean office fosters a data-driven culture where data is central to decision-making. Lean office's focus on efficiency complements this by encouraging employees to use AI insights to refine processes and create a culture of a workforce that is confident and proficient in using data to make decisions.

The integration of Lean office with AI also encourages transparency, open communications, and collaboration, by breaking down functional silos by providing shared platforms for knowledge management, information, and insight dissemination.

A Lean office-AI culture is more agile with an ability to adapt quickly to changes in the market or industry by leveraging AI's predictive capabilities and Lean office's iterative improvement processes. This results in improved efficiency, reduced cycle times, lower costs, improved overall operational efficiency, and better quality of products and services. Quality improvement is a result of AI's ability to provide proactive, precise, data-driven insights into process performance, leading to higher-quality outputs.

Lean office-AI results in sustainable growth by fostering a culture of continuous improvement, efficiency, and a competitive edge in an increasingly AI-driven marketplace. Thus, integrating Lean office with AI not only optimizes operations but also cultivates a progressive workplace culture that embraces change and innovation, driving superior business results and providing a competitive edge to organizations.

INTEGRATING LEAN IN OFFICE WITHIN THE LEAN-AI MODEL

Integration of AI with the Lean office methodology can be accomplished at a strategic, operational, and tactical level to maximize its impact as seen below:

STRATEGIC LEVEL

At the strategic level, the focus is on aligning AI with the organization's strategic plans, long-term goals, and the broader Lean office methodology to drive culture change, sustainable growth, and innovation across the organization, supply chain, and customers. At the strategic levels, AI is integrated and aligned with the strategic plan, vision, leadership, culture, and resource allocation.

When using Hoshin Kanri or other strategy creation tools, it is important to include AI along with a Lean office strategy to eliminate waste, improve flow, and enhance quality across the organization. This might include a strategy to invest in the predictive analytics capability of AI for customer behavior, supply chain optimization, or automating strategic decision-making processes. AI should be integrated into the organization's vision for Lean office by identifying key areas where AI can support long-term goals, such as enhancing customer value, driving innovation, and increasing competitiveness in the market.

It can also help to cultivate a leadership culture that embraces AI as a tool for continuous improvement. Leaders should be trained to understand AI's capabilities and to foster a data-driven mindset across the organization, to ensure alignment between AI and Lean office initiatives. Another strategic application of Lean office-AI is to allocate resources strategically for AI development and integration, focusing on areas that offer the highest return on investment. This might include automating complex data analysis tasks that drive Lean office improvements or investing in AI tools that enhance the strategic planning effort and its deployment across the organization.

OPERATIONAL LEVEL

At the operational level, the focus is on optimizing core business processes by integrating AI tools that enhance Lean office practices, such as waste reduction, process standardization, and continuous improvement using process optimization, advanced problem-solving, data-driven decision-making, and operational agility.

Process optimization, such as workflow automation, supply chain management, and customer service, can be achieved using AI that can analyze and optimize key office processes. For example, AI-driven predictive analytics can forecast demand, allowing for more efficient resource allocation and reducing overproduction (a form of waste in Lean office). Advanced problem-solving can result from implementing AI tools that support Lean office's problem-solving methodologies, such as Six Sigma and Kaizen. AI can analyze large datasets and conduct root cause analysis of inefficiencies or predict future issues before they arise, enabling more proactive management at an operational level.

Data-driven decision-making in the Lean office can be enhanced by integrating AI by providing real-time data and insights. For example, AI can automate data collection and analysis, providing actionable insights that align with Lean's focus on fact-based decisions. The agility of operations can be enhanced by using AI, resulting in rapid adjustments to processes in response to fluctuations in demand, market conditions, or internal challenges. This could involve AI tools that automatically adjust staffing levels or inventory based on real-time data.

TACTICAL LEVEL

At the tactical level, AI can be integrated into day-to-day activities to support Lean office practices in areas such as task automation, continuous improvement, and employee empowerment. Automating repetitive, low-value tasks using AI can free up employees to focus on more strategic, value-added activities. For example, AI can handle routine administrative tasks, such as data entry or scheduling, reducing wasted time and improving efficiency. AI tools can support continuous improvement efforts by providing ongoing monitoring and feedback. For instance, AI can analyze performance metrics in real time, identifying areas for improvement, and automatically suggesting adjustments to optimize the processes.

AI tools can assist in empowering employees in decision-making and problem-solving. AI can offer recommendations based on data analysis, helping employees make more informed decisions that align with the principles of the Lean office. AI can ensure consistency in standard work practices. AI-driven systems can monitor adherence to Lean office standards, alerting managers to various deviations and even recommending corrective actions to maintain quality and efficiency.

As seen in Figure 12.1, integrating AI within the Lean office methodology at strategic, operational, and tactical levels enables organizations to fully leverage the benefits of both AI and Lean. This integrated approach ensures that AI becomes

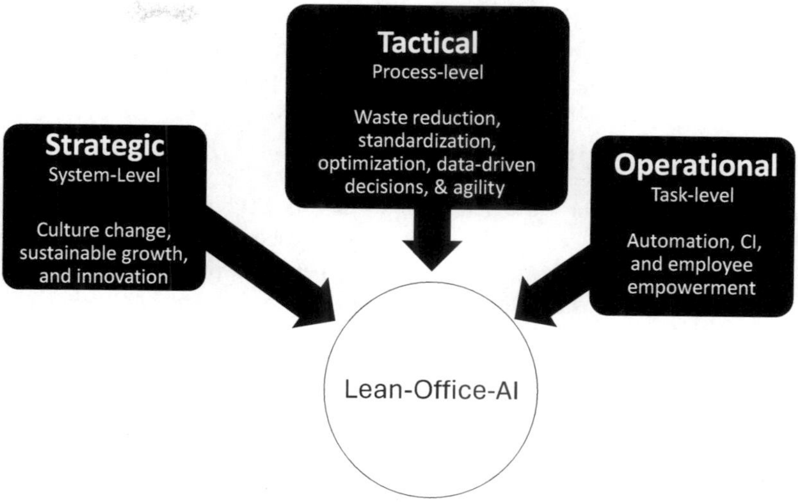

FIGURE 12.1 Lean office-AI model.

a powerful enabler of Lean principles, transforming the office environment into a highly efficient, innovative, and agile space.

LEAN OFFICE-AI DATA PRIVACY AND ETHICS CONSIDERATIONS

Integrating Lean office with AI demands careful attention to data privacy and ethical considerations, as these elements play a pivotal role in building trust and safeguarding both organizational and employee interests. Here are the key considerations:

DATA PRIVACY AND CONFIDENTIALITY

Employee Data Protection: AI tools often rely on extensive data, including employee performance metrics and behavioral insights. Protecting this sensitive data is essential to avoid potential misuse or unintended consequences.

Customer Information Security: If customer data is involved, particularly in industries with high data sensitivity like healthcare or finance, robust safeguards like data encryption, anonymization, and secure access controls must be in place to ensure compliance with standards like GDPR and HIPAA.

Data Minimization: In line with Lean principles, only necessary data should be collected to avoid data bloat and minimize risks. This reduces exposure and helps maintain focus on efficiency.

TRANSPARENCY AND ACCOUNTABILITY

AI Algorithm Transparency: Lean office processes prioritize efficiency and waste reduction, but when AI is involved, it's important to clarify how decisions are made. Employees need to understand how data influences AI-driven recommendations or decisions that affect their roles or productivity evaluations.

Bias and Fairness: AI models can inadvertently introduce or reinforce biases, impacting hiring, performance, or advancement decisions. Regular bias audits and diverse training data help ensure fairness in processes influenced by AI.

CONSENT AND EMPLOYEE AUTONOMY

Informed Consent for Data Use: Employees should be informed about how their data is collected, analyzed, and used, especially when AI tools are involved in productivity monitoring or personal performance insights. Informed consent is crucial for maintaining trust and ethical standards.

Balancing AI Oversight with Human Judgment: While AI can optimize Lean processes, maintaining a balance where human judgment prevails in ethical or sensitive decisions is essential to avoid over-reliance on algorithms.

ETHICAL USE OF INSIGHTS AND OUTCOMES

Avoiding Micromanagement: Lean office promotes efficiency without compromising respect for the individual. AI insights should not lead to invasive micromanagement but instead support employees in optimizing their own performance autonomously.

Alignment with Organizational Values: AI-driven Lean office transformations should align with the organization's values, ensuring the technology is used to support, rather than exploit, the workforce.

COMPLIANCE AND REGULATORY ADHERENCE

Adherence to Data Protection Regulations: Compliance with global and local regulations is fundamental, especially when handling sensitive employee or customer data. Regular reviews and updates ensure that Lean-AI implementations remain legally sound.

Proactive Risk Assessments: Regular assessments identify potential privacy risks, allowing organizations to adapt processes before issues arise. This proactive approach aligns with Lean's focus on continuous improvement and risk mitigation.

In summary, Integrating AI with Lean office can streamline processes and enhance productivity, but it must be implemented with a commitment to ethical standards and data protection. By establishing clear policies, maintaining transparency, and respecting employee autonomy, organizations can foster a culture that supports both innovation and trust.

LEAN OFFICE-AI INTEGRATION INDUSTRY EXAMPLES

Integrating Lean office with AI in various industries has led to transformative improvements in efficiency, quality, and overall performance (Bortolotti, Thomas & Romano, Pietro & Nicoletti, Bernardo et al., 2009). Here are some notable industry examples:

Healthcare–Patient Flow: Implementation of Lean office methodologies to streamline patient flow, reduce wait times, and eliminate waste in administrative processes. Utilizing AI-driven predictive analytics to forecast patient admissions, optimize staff scheduling, and personalize patient care plans. This can result in enhanced patient satisfaction through reduced wait times, improved resource allocation, and higher quality of care. The combination of Lean and AI can lead to a more responsive and efficient healthcare delivery system (Graban, 2016).

Healthcare-Waste Reduction: Adoption of Lean office methodology to improve administrative and supply chain processes, reducing waste in areas like order processing and resource allocation. AI-powered predictive analytics tools can be introduced to anticipate equipment maintenance, manage inventory, and optimize staffing levels. This can help with predicting equipment failures and reducing downtime while streamlining inventory and staff management. Empowered staff can use data insights to make better decisions and continually improve workflows. This can result in a significant reduction in operational costs and faster response times for equipment issues, leading to improved patient care and satisfaction.

Finance: Application of Lean office principles to improve back-office operations, reduce processing times, and eliminate inefficiencies in workflows. Leveraging AI for fraud detection, risk assessment, and automated customer service through chatbots can result in enhanced operational efficiency, reduced error rates, and more personalized customer experiences. The integration can facilitate faster decision-making and better risk management.

Banking and Financial Services: Implementation of Lean to streamline back-office processes, eliminating unnecessary steps in tasks such as loan processing and customer service. Introducing AI tools for automating routine tasks like document analysis, fraud detection, and customer service through chatbots can free up employees from routine administrative tasks, allowing them to focus on complex, value-added services. AI-driven document scanning and processing can significantly improve efficiency, reducing human error and delays. This can result in improved customer satisfaction and a reduction in processing times for loans and other transactions (George, 2003).

Insurance: Adoption of Lean office to improve claims processing and customer service functions, eliminating non-value-added steps in these processes. AI-powered automation tools can be introduced for tasks like claims triage, fraud detection, and policy recommendations, resulting in significantly reduced claims processing time with AI tools that automatically assess damage and recommend settlements based on historical data. It can also allow the claims handlers to focus on complex cases that require human judgment, while AI can automate less complex claims. This can result in significant improvements in processing speed, reduced fraud, and enhanced customer satisfaction.

Retail – Supply Chain: Employing Lean office strategies to optimize inventory management, streamline supply chains, and enhance store operations. Adoption of AI for demand forecasting, personalized marketing, and automated checkout systems. This can result in improved inventory accuracy, personalized customer experiences, and accelerated checkout processes. This integration can lead to higher sales, reduced costs, and increased customer loyalty.

Retail-Waste Reduction: Adoption of Lean office principles in logistics and administrative operations to minimize waste, including in customer service, HR, and vendor management. Integrating AI systems, such as NLP and machine learning, can be introduced to automate customer inquiries, improve inventory management, and optimize shipping routes. This can result in the ability of customer service representatives to handle complex issues, as chatbots manage simpler queries, enhancing job satisfaction and efficiency. AI-driven inventory management and shipping optimizations can complement Lean office's focus on reducing waste and improving flow. This can also result in reduced operational costs, faster customer query resolutions, and shorter delivery times.

Information Technology (Agile): Implementation of Lean software development practices to accelerate project delivery, improve code quality, and enhance team collaboration. Integrating with AI for code analysis, automated testing, and intelligent project management tools. This can result in increased development speed, reduced bugs, and foster a culture of continuous improvement. The combination can enable more innovative and reliable software solutions (Poppendieck & Cusumano, 2012).

Telecommunications: Applying Lean office methodologies to improve network operations, reduce service disruptions, and enhance customer support processes. Integrating this with AI for network optimization, predictive maintenance, and automated customer service through virtual assistants can result in enhanced network reliability, reduced maintenance costs, and faster, more efficient customer support. The synergy between Lean and AI can contribute to a more robust and customer-centric service offering.

Logistics and Supply Chain: Implementing Lean office principles to optimize warehouse operations, reduce lead times, and eliminate bottlenecks in the supply chain. Integrating this with AI and deploying AI for route optimization, demand forecasting, and automated sorting systems can result in achieving faster delivery times, increased operational efficiency, and minimized logistical costs. The integration facilitates a more agile and responsive supply chain.

Document Control: Adoption of demand flow in the office techniques to reduce cycle times, minimize waste, optimize processes, plan for required resources, and enhance product quality. Integrating and deploying AI for simulating the daily volume of documents to be updated, the type of documents to be updated, blockages, quality control, and change control signature process optimization. This can result in significant reductions in cycle times, better resource planning and allocation, reduction of blockages, improved quality, and optimized cycle time for signatures needed for change control of documents. This synergy between Lean and AI can foster a more streamlined, resilient, and agile document control environment.

In these examples, integrating Lean principles with AI technologies can enhance productivity, employee engagement, and business outcomes, showing how AI can magnify the benefits of Lean in various office settings. Several industries have already successfully integrated Lean principles with AI in office environments, streamlining and transforming their processes and boosting efficiency.

LEAN OFFICE-AI ADVANTAGES

Integrating Lean office practices with AI offers a powerful combination that enhances efficiency, productivity, and innovation. Here are some key advantages of this integration:

Improved Decision-Making: AI provides real-time analytics and predictive insights, allowing for more informed decision-making. This aligns with Lean office's emphasis on data-driven management, helping organizations make better decisions that improve operational efficiency and customer satisfaction. AI's predictive capabilities enable organizations to anticipate issues before they arise. By integrating AI with Lean office's continuous improvement practices, potential problems can be addressed proactively, reducing downtime and enhancing process stability.

Better Quality and Consistency: AI-driven systems can monitor processes continuously, ensuring that quality standards are met consistently. This supports the Lean office's focus on maintaining high quality by automatically detecting and correcting deviations from standard procedures. AI tools can help enforce standard work practices across the organization, ensuring consistency and reducing variability. This leads to more reliable outcomes and higher customer satisfaction.

Enhanced Efficiency and Productivity: AI automates routine and repetitive tasks such as data entry, scheduling, and document management, freeing up employees to focus on more strategic activities. This complements Lean office's goal of reducing non-value-added activities, resulting in increased productivity. AI enhances Lean office's process optimization efforts by identifying inefficiencies and bottlenecks more accurately. With AI's ability to analyze vast amounts of data quickly, processes can be streamlined more effectively, leading to faster cycle times and reduced waste.

Accelerated Continuous Improvement: AI can continuously monitor key performance indicators (KPIs) and provide insights for ongoing improvement. This accelerates the Lean office process of continuous improvement (office kaizen) by offering timely data that can be acted upon immediately. AI integration creates faster feedback loops by analyzing data in real time and suggesting improvements instantly. This ensures that Lean office's iterative processes are more dynamic and responsive to changes, leading to quicker and more effective enhancements.

Empowered Workforce: By automating mundane tasks, AI allows employees to engage in more meaningful work, fostering a culture of innovation and learning. This complements Lean office's emphasis on empowering employees and encourages continuous skill development. AI tools provide employees with decision support, enhancing their ability to solve problems and contribute to process improvements. This empowerment aligns with Lean office's focus on involving all employees in the pursuit of excellence.

Cost Reduction: AI reduces operational costs by automating tasks, optimizing resource allocation, and minimizing errors. Combined with Lean office's focus on eliminating waste, this results in significant cost savings. AI helps organizations optimize the use of resources such as labor, materials, and time. This leads to a leaner operation with lower overheads and better financial performance.

Greater Agility and Responsiveness: AI enables organizations to adapt quickly to market changes, customer demands, and operational challenges. This agility is enhanced by Lean's flexible processes, allowing for quick pivots and adjustments. The integration of AI with Lean practices makes it easier to scale operations without compromising on efficiency or quality. This is particularly beneficial in dynamic markets where rapid growth or downsizing is required.

Competitive Advantage: The combination of AI's cutting-edge technology with Lean's proven methodologies drives innovation, giving organizations a competitive edge. AI can uncover new opportunities for improvement and innovation that might be missed using traditional Lean techniques alone. With improved processes, higher quality, and faster response times, organizations can better meet customer needs, leading to increased customer satisfaction and loyalty.

In summary, the integration of Lean office and AI is a strategic advantage that enhances efficiency, decision-making, quality, continuous improvement, and employee empowerment while reducing costs and increasing agility. This powerful combination not only drives operational excellence but also positions organizations to thrive in a rapidly evolving business landscape (Davenport & Bean, 2018).

AGILE-AI ADVANTAGES

Enhanced Decision-Making and Data-Driven Insights: The integration of Agile and AI enhances decision-making by leveraging AI's data analysis capabilities to provide real-time insights, predictions, and recommendations that improve project outcomes (Cha, Lee, & Kim, 2018).

Increased Efficiency and Automation: Agile practices combined with AI can automate repetitive tasks, streamline workflows, and reduce manual effort, resulting in increased efficiency and faster delivery cycles (He, Zhang, & Liu, 2020).

Improved Adaptability and Flexibility: Advantage: Integrating AI with Agile methodologies allows for greater adaptability and flexibility in responding to changing requirements and emerging trends by providing predictive analytics and automated adjustments (Wirtz, & Müller, 2020).

These advantages highlight how integrating Agile with AI can enhance decision-making, increase efficiency, and improve project management in software development.

CONCLUSION

Lean is considered a universal tool to support organizations in manufacturing and in the office environment for optimizing business processes. Integration of Lean office practices with AI offers a transformative approach to modern organizational efficiency, quality, and innovation. By combining the structured, waste-reducing methodologies of Lean with the data-driven, automation capabilities of AI, organizations can realize substantial improvements in both operational effectiveness and strategic agility.

One of the primary advantages of this integration is the enhancement of decision-making processes. AI's real-time analytics and predictive insights align seamlessly with Lean's emphasis on data-driven management, enabling organizations to make informed decisions that enhance operational efficiency and customer satisfaction. AI's ability to anticipate potential issues before they arise allows for proactive problem-solving, which complements Lean's focus on continuous improvement and process stability.

Quality and consistency are further improved through AI-driven systems that continuously monitor processes, ensuring adherence to established standards. This automation reduces variability, upholds quality benchmarks, and supports Lean's goal of maintaining high standards across operations. By enforcing standard work practices through AI, organizations achieve more reliable outcomes, leading to higher customer satisfaction and a stronger competitive edge.

EXERCISES

1. What are the current system and process limitations of your Lean office approach, and how can you apply AI to overcome these limitations?

Lean Office Step	System and Process Limitations	AI Approach to Overcome System and Process Limitations
6S		
VSM/Swim Lane Mapping		
7 Wastes		
Poka-Yoke		
Lean office Kaizen		

2. What are the current cultural limitations of your SMED approach, and how can you apply AI to overcome these limitations?

SMED Step	Cultural Limitations	AI Approach to Overcome Cultural Limitations
Step 1: Determine the setup operation		
Step 2: Prepare for changeover		
Step 3: Changeover internal steps		
Step 4: Perform calibrations		
Step 5: Run trials and adjust		

3. What are some short- and long-term benefits to your department and organization for integrating AI into Lean office?

#	Short-Term Benefits	Long-Term Benefits
Example	*Decreased hand-off's*	*Improved process quality*
1		
2		
3		
4		
5		
6		

4. Reflect on the ethical implications of using AI in Lean office within your organization. Consider aspects such as employee privacy, job displacement, and data security. Develop strategies to mitigate the negative aspects of the workforce.

#	Ethical Considerations	Strategies to Mitigate
Example	*Job displacement fears*	• *Provide upskilling and reskilling programs* • *Clear communication about AI's role*
1		
2		
3		
4		
5		
6		

5. What are the technology considerations for integrating AI into your Lean office framework within your organization? Develop strategies for the necessary technology integration.

#	Technology Necessary for AI	Strategy for Integration
Example	*Real-time data analytics*	• *Develop robust data infrastructure* • *Train staff on data management*
1		
2		
3		
4		
5		
6		

REFERENCES

Beck, K., et al. (2001). *Agile alliance*. http://agilemanifesto.org/

Bhamu, J., & Sangwan, K. S. (2014). Lean manufacturing: Literature review and research issues. *International Journal of Operations & Production Management*, *34*(7), 876–940.

Boehm, B. W. (2006). *Get ready for agile methods, with care*. IEEE Computer Society.

Bortolotti, T., Romano, P., & Nicoletti, B. (2009). Lean first, then automate: An integrated model for process improvement in pure service-providing companies. *IFIP Advances in Information and Communication Technology*, *338*, 579–586. https://doi.org/10.1007/978-3-642-16358-6_72

Cha, J., Lee, K., & Kim, J. (2018). Artificial intelligence and its applications in the agile development process. *Journal of Software Engineering and Applications*, *11*(6), 223–238.

Davenport, T. H., & Bean, R. (2018). *The AI advantage: How to put the artificial intelligence revolution to work*. MIT Press.

George, M. L. (2003). *Lean Six Sigma for service: How to use lean speed and six sigma quality to improve services and transactions*. McGraw-Hill.

Graban, M. (2016). *Lean hospitals: Improving quality, patient safety, and employee engagement* (3rd ed.). CRC Press.

He, L., Zhang, Y., & Liu, X. (2020). Leveraging artificial intelligence to enhance agile software development: A framework and case study. *Journal of Systems and Software*, *166*, 110573.

Hughes, B., & Cotterell, M. (2009). *Software project management* (5th ed.). McGraw-Hill Education.

Leffingwell, D. (2011). *Scaling software agility: Best practices for large enterprises*. Addison-Wesley.

Netland, T. H., & Ferdows, K. (2014). The S-curve effect of lean implementation. *Production and Operations Management*, *23*(8), 1279–1299.

Papadopoulos, T. (2010). *Lean administration: Case studies in leadership and improvement*. Taylor & Francis.

Poppendieck, M., & Cusumano, M. A. (2012). *Lean software development: Applying agile principles to your organization*. Addison-Wesley.

Poppendieck, M., & Poppendieck, T. (2003). *Lean software development: An agile toolkit*. Boston: Addison-Wesley.

Tapping, D., & Dunn, A. (2006). *Lean office: Demystified II* (2nd ed.). MCS Media, Inc.

Wirtz, B. W., & Müller, W. (2020). AI-enabled agility: The integration of AI into agile management practices. *Business Horizons*, *63*(6).

Womack, J. P., Jones, D. T., & Roos, D. (2007). *The machine that changed the world*. Simon & Schuster.

13 Kaizen and Kaikaku-AI

Zeeshan Zia

AI FOR PROCESS IMPROVEMENT

AI can impact process improvement using kaizen and kaikaku at strategic, tactical, and operational levels as seen in Figure 13.1. We draw inspiration from the exponential improvements in the quality and quantity of software products over the past several decades. Key vectors for this improvement are the abilities of software developers to rapidly search for ideas (e.g. Stack Overflow and open source) as well as to copy-paste and build on top of existing implementations (e.g. due to Object-Oriented Programming, software frameworks, and GitHub).

Unfortunately, there are no such mechanisms available in manufacturing engineering for sharing and building on top of existing ideas and infrastructure. Manufacturing and industrial engineers always start from scratch when designing and implementing a new process. There don't even exist efficient mechanisms to transfer ideas within an organization. A brilliant industrial engineer who comes up with an efficient process for a "car door assembly" process can hardly enforce compliance with their standard process across shifts within their own facility, let alone share that process across multiple factories of the same organization. Typically, those process learnings get buried in an Excel spreadsheet, and none of the other factories ever learn about it, as it's just one of hundreds of such processes and there's always short-term work that such engineers have to contend with. An industry that loses 20% of revenues to quality issues (Faciane, 2018), and that is expected to lose $1 trillion (Egan, 2021) a year due to labor issues cannot afford these types of information processing and sharing inefficiencies.

Now, in software, the next level of productivity has been unlocked with the Copilot revolution which is allowing software engineers and stakeholders to obtain immediate productivity boosts of 40% (Maxwell, 2023). Our vision is to leverage multi-modal foundation models trained on manufacturing data to skip several steps ahead and build such a Copilot for manufacturing that will inject software-like dynamics in manufacturing process design and implementation. This will dramatically accelerate innovation in process design and massively increase productivity, quality, and comfort in process implementation.

Our approach is to deeply understand industrial process design and implementation workflows, build simple GUIs to model these workflows, and use Generative AI to augment the human judgment involved both on the process design and process implementation fronts. To achieve this, we are building a multi-modal foundation model specific to manufacturing assembly. We have added the ability of very fine-grained manufacturing assembly-related activity understanding and part defect understanding from video and sensor data to a pre-trained LLM backbone

 DOI: 10.1201/9781003532972-13

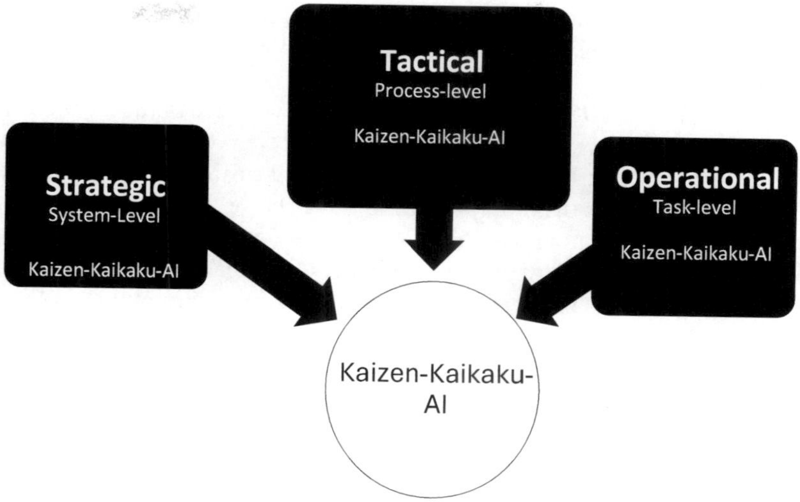

FIGURE 13.1 Kaizen-kaikaku-AI model.

(specifically Llama3 from Meta). We are training this model on customer data and other forms of proprietary data that only we have access to, together with human annotations that our customers are providing.

Further, since manufacturing deals more with atoms than bits, we embody the aforementioned insights by connecting to camera and sensor streams, enabling the AI to act through PLCs and actuators. Today this takes the form of real-time worker guidance, just-in-time training, quality-related alerts, and physical mistake-proofing such as stopping a conveyor belt when critical issues are detected. In the future, this will expand to assisting the engineer in rapidly identifying where automation will truly be effective and then automatically generating such automation, for example, generate programs for cobots to partly and eventually fully automate processes.

TIME STUDY AND STATION IMPROVEMENT

The integration of AI into time study and station improvement has revolutionized the way manufacturing processes are analyzed and optimized. Traditional time and motion studies require significant human effort to manually observe, document, and break down work cycles into discrete steps, but AI-enabled tools like Kaizen Copilot make this process more streamlined and efficient.

AI-based time study leverages video analysis technology to automate the segmentation of work cycles, which provides a more precise and detailed understanding of the various tasks that make up an assembly process. The process typically begins with a simple video recording of an assembly cycle, which can be captured using any standard camera or smartphone. Once the video is uploaded into the software, users can define key elements like the start and end of the cycle, the work region, and the number of steps within the process. The AI then automatically identifies and breaks down the cycle into meaningful steps, eliminating the need for extensive manual annotations.

By focusing on each step individually, users can review and refine the breakdown. Kaizen Copilot allows users to replay each step and even rename or adjust steps as needed, ensuring that the analysis is both accurate and customized to specific needs. The AI further enhances the process by asking suggestive yes/no questions for each identified step, incorporating user feedback into its final improvement recommendations. This interaction between human input and AI-generated suggestions helps identify potential inefficiencies and recommends actionable improvements.

One of the key benefits of AI-based time study is its ability to provide real-time, data-driven insights. For example, Kaizen Copilot can estimate potential reductions in cycle time and calculate the resulting cost savings. The software also analyzes operator motions, identifying areas where movements may be haphazard or inefficient. It even provides ergonomic suggestions by dividing the workspace into different zones and measuring activity levels within each zone. By optimizing the layout and motion paths, manufacturers can reduce unnecessary motions and create more efficient workflows.

In addition to time studies, the AI performs station audits, which include ergonomic assessments and standard work adherence evaluations. The software generates audit questions related to potential inefficiencies and non-value-adding activities. This allows users to pinpoint areas of waste, such as unnecessary movement or poor sequencing of steps, which can then be addressed to further enhance productivity.

At the end of the process, Kaizen Copilot produces detailed reports, including standard work charts, spaghetti diagrams of worker movements, and other visualizations that allow for an in-depth understanding of the process. All these analytics can be exported for integration into existing Kaizen procedures within an organization.

The AI-based time study and station improvement module is not just a tool for analyzing the present; it actively contributes to continuous improvement by allowing users to record feedback on the effectiveness of the suggested changes. This ongoing feedback loop enables the AI to learn and become more accurate over time, ensuring that future recommendations are even more tailored to specific processes and needs.

In conclusion, AI-based tools for time study and station improvement, like Kaizen Copilot, offer manufacturers a powerful way to boost efficiency, reduce costs, and enhance ergonomic safety by automating the traditionally labor-intensive process of cycle analysis. The system's ability to provide actionable insights and recommendations empowers companies to make data-driven decisions that lead to substantial improvements in productivity and operator well-being.

AUTOMATIC LINE BALANCE

AI-based automatic line balancing is a transformative approach to optimizing production workflows. Traditionally, line balancing required extensive manual calculations and iterative adjustments to distribute tasks evenly across stations, resulting in a balanced workflow. However, AI-enabled tools, like the automatic line balancing module in Kaizen Copilot, streamline this complex process, enabling manufacturers to analyze and adjust production lines quickly and with precision.

Line balancing is the process of assigning tasks to different workstations in such a way that the entire production line operates in harmony, achieving high levels of efficiency while avoiding bottlenecks. The key objective is to ensure that the cycle time for each station is aligned with the takt time – the rhythm at which products must be produced to meet customer demand. An imbalanced line leads to inefficiencies, including idle time, overburdening of specific stations, and suboptimal utilization of labor and equipment.

Kaizen Copilot's line balancing tool offers a user-friendly interface that automates much of this process. Industrial engineers can start by importing station-level time data, either from Kaizen Copilot's own station design tool – which uses video-based time studies – or by manually entering time estimates. With this data in place, the AI system runs multiple iterations of potential work distributions in the background, taking into account constraints such as the number of operators, desired cycle times, and production volume targets. This rapid computational power allows engineers to quickly assess different scenarios, compare options, and select the most efficient line configuration within seconds.

A distinguishing feature of Kaizen Copilot's line balancing tool is its ability to visually present complex line structures using various charts and diagrams given next.

PRECEDENCE DIAGRAM

This visual map represents the logical sequence and dependencies between tasks, illustrating how each task is connected to others. This allows engineers to quickly understand the flow of operations and identify potential chokepoints in the production line.

LINE BALANCE CHART

This chart highlights bottleneck stations and displays the cycle time and utility percentage for each station, making it easy to see where adjustments are needed. It also provides a summary of the overall line efficiency, helping engineers evaluate whether the production line is operating optimally.

YAMAZUMI CHART

A stacked bar chart that breaks down the distribution of work elements across different operators or stations. It categorizes tasks into value-added and non-value-added activities, enabling engineers to visualize how work is allocated and where efficiency can be improved.

With these visual aids, engineers can gain a comprehensive understanding of the current state of the line and identify areas for improvement. Moreover, Kaizen Copilot's AI-powered suggestions provide automated recommendations for optimizing work allocation. For example, if a specified takt time of 65 seconds results in an efficiency of 77.5% with five operators, the AI might suggest a slight adjustment – reducing the takt time to 61 seconds – to achieve a higher efficiency of 86% without increasing headcount. These suggestions are based on detailed analyses of task dependencies and cycle times, helping engineers make data-driven decisions that lead to significant productivity gains and cost reductions.

Another valuable feature of AI-based line balancing is its ability to perform rapid rebalancing when conditions change. If a new takt time is introduced, or the number of operators fluctuates, users can simply update the parameters and let the tool generate a new, optimized line balance scenario. This flexibility ensures that the production line can adapt dynamically to varying production demands or workforce availability without the need for extensive recalculations or manual reconfiguration.

Additionally, Kaizen Copilot allows users to explore cost-saving opportunities by evaluating different line-balancing scenarios. For example, running a line at a faster takt time may reduce the total number of operators needed, leading to potential savings in labor costs. Alternatively, balancing a line to reduce bottlenecks may enable the line to meet production targets more consistently, eliminating the need for overtime or additional shifts.

Once the optimal line balance is determined, Kaizen Copilot generates comprehensive reports that include all the key diagrams and data tables, such as step allocation for each station, their respective cycle times, and suggested improvements. These reports can be downloaded in PDF or Excel formats and shared with production teams for easy implementation on the shop floor.

In summary, AI-based automatic line balancing, as provided by Kaizen Copilot, empowers manufacturing engineers to optimize production lines quickly and accurately. By automating the traditionally manual process of evaluating and adjusting task allocation, these tools provide actionable insights that enhance line efficiency, reduce bottlenecks, and lower operational costs. The flexibility to rapidly adapt to changes in takt time, workforce, or production volume makes AI-based line balancing a powerful asset in any modern manufacturing environment.

Case Study: Automotive Components Assembly Line

This case study demonstrates the reduction of operator headcount from six to four on an automotive part sub-assembly line by providing alternative line balance within two hours of effort including data recording.

Some challenges included time study and line balancing take two to three weeks of manufacturing engineer's time, which bottlenecks their operational excellence strategy.

The immediate impact was that the Lean engineer temporarily mounted a smartphone above each of the six stations on the line one by one, capturing 10–15 minutes of assembly video from each station (20–25 cycles). Kaizen Copilot was able to automatically break down the cycles into steps without requiring detailed annotation; followed by several automatic line balance recommendations. The system recommended a balance with just four workers. The overall work took just two hours of effort from the Lean engineer.

The engineer further printed the SOP from the tool and recommended the new SOP to the line supervisor, who was able to validate and implement it the following week.

ERGONOMICS ANALYSIS AND RECOMMENDATIONS

Ergonomics analysis is a critical aspect of optimizing work environments in manufacturing and other industrial settings. Poor ergonomic practices can lead to musculoskeletal disorders (MSDs), lower productivity, and increased injury rates. To address these challenges, AI-based tools like Kaizen Copilot provide comprehensive ergonomics analysis and improvement recommendations that can help reduce risks associated with repetitive motion, awkward postures, and excessive force during manual tasks.

Kaizen Copilot's ergonomics suite leverages advanced computer vision to analyze video recordings of workers performing tasks and generates detailed insights about postural risks. The platform can process videos recorded from ordinary cameras, extract precise 3D skeletal poses, and calculate joint angles. These capabilities allow non-ergonomists to assess and improve workstation designs, task sequences, and workflows without the need for extensive manual observation or expertise.

REBA AND RULA ASSESSMENTS

The Rapid Entire Body Assessment (REBA) and Rapid Upper Limb Assessment (RULA) are two key methodologies integrated into Kaizen Copilot's ergonomics module. These tools assess the risk of MSDs by evaluating postures, forces, and repetition during tasks.

REBA evaluates whole-body posture, including the exertion of forces in lifting, pushing, pulling, and carrying tasks. It is particularly useful for tasks that involve standing or movement across an entire workstation.

RULA focuses on upper body posture, such as arms, wrists, and neck, and is more applicable to benchtop or seated work environments.

The AI automatically detects intervals in the video where risky postures are present according to REBA or RULA guidelines. It identifies limbs at risk and provides severity ratings using color-coded body charts. This intuitive visual feedback allows users to quickly understand which body parts are under the most strain and require ergonomic improvements.

Kaizen Copilot further enhances this analysis by offering time-based and limb-specific insights:

Time-based analysis shows the risk throughout a task cycle using color-coded bar graphs, highlighting periods of low, medium, and high risk.

Limb analysis tracks joint angles over time and compares them to ergonomic thresholds, enabling users to identify high-risk positions and plan for adequate rest intervals between tasks. By optimizing work tasks and equipment placement, these insights promote better posture and reduce the likelihood of injury.

NIOSH LIFTING EQUATION

Manual material handling tasks, such as lifting and lowering, are common sources of ergonomic risk. The NIOSH lifting equation helps assess the safety of these tasks by calculating a recommended weight limit that workers can lift without increasing their risk of lower back pain.

Kaizen Copilot's AI simplifies this calculation by automatically detecting critical parameters such as vertical and horizontal lift distances from the video footage. Users can input additional variables, such as the frequency of lifts or the quality of the handles on objects, to refine the analysis.

The system classifies lifts based on factors like height and power zone, providing a time series chart that helps pinpoint unsafe repetitions and critical lifting times. This detailed analysis helps in identifying hazardous lifting practices and making adjustments to mitigate injury risks.

SNOOK TABLE ANALYSIS

The Snook Table is another tool integrated into Kaizen Copilot, used to evaluate the ergonomic suitability of manual material handling tasks. It provides guidelines for recommended heights of work surfaces, seating, and equipment, based on anthropometric data.

By leveraging AI object tracking and pose estimation, Kaizen Copilot automates the assessment of tasks against Snook Table recommendations. The system calculates the lift's vertical and horizontal positions, evaluates task safety, and assigns a color-coded risk index to illustrate physical stress levels. If the task exceeds the acceptable threshold, the system flags it as a significant injury risk, prompting users to consider redesigning the workstation or modifying the task to fit within safe parameters.

HAND GRIP AND INSERTION FORCES

Tasks that involve repetitive gripping or forceful insertions can lead to issues such as carpal tunnel syndrome and other hand and wrist injuries. Kaizen Copilot's ergonomics module includes an assessment for hand grip and insertion forces. The system uses AI to detect different grip types (e.g., lateral pinch, power grip) and wrist postures in video recordings, then evaluates the cumulative frequency and force exerted during these tasks.

By comparing the actual forces to recommended limits, Kaizen Copilot identifies whether the grip forces exceed safe thresholds. The AI provides feedback in the form of color-coded indicators, showing where excessive force is being applied and recommending adjustments to reduce strain.

COMPREHENSIVE REPORTS AND RECOMMENDATIONS

After performing ergonomic assessments, Kaizen Copilot generates comprehensive reports that summarize key findings and recommendations. These reports include detailed charts, graphs, and risk assessments that highlight critical ergonomic issues and provide actionable insights for improvement. Users can download these reports in PDF format to share with teams and implement changes on the shop floor.

Overall, Kaizen Copilot's AI-enabled ergonomics tools provide an efficient, scalable way for manufacturers to assess and improve workplace safety. By automating complex ergonomic evaluations and providing data-driven recommendations, the platform helps organizations minimize risks, enhance worker well-being, and improve productivity.

Case Study: Automotive OEM IP Hood Garnish Assembly Line

This case study focuses on a material handling process, in which Kaizen Copilot was able to: (1) eliminate 12 monthly defects, (2) reduce tray handling from 24 times per cycle to 12 times per cycle, (3) eliminate downtime of 16 minutes daily, and (4) eliminate four annual discomfort cases and one annual ergonomic incident (MSD) worth $21k, by highlighting issue and suggesting to implement a One Touch Dolly and Flow Rack system.

The challenge was that the team members were experiencing ergonomic overburden at Automotive OEM IP Hood Garnish Process. This resulted in 12 defects per month with a potential to flow out to customers, and 16 minutes of daily downtime (8 minutes of short time to IP line daily × 2). Four discomfort cases were being opened a year due to ergonomic overburden and 1 MSD incident costing $20,837.

The immediate impact was that the video recording of the process was captured with a smartphone, and uploaded to Kaizen Copilot, which automatically assessed the situation, measured heights, distances, and estimated lift forces; and also recommended the implementation of a "One touch Dolly" and "flow rack system." The implementation of this system eliminated all issues identified. Dolly allows the full and empty trays to be loaded/unloaded simultaneously. Reduced overall process time, which allowed team members to confirm quality. Eliminated downtime (16 minutes daily). No cases of ergonomic overburden since implementation.

QUALITY PLANNING: P-FMEA AND CONTROL PLANS

Effective quality planning is essential for ensuring product reliability, preventing defects, and minimizing risks in manufacturing processes. AI-based tools, like those integrated into Kaizen Copilot, enhance traditional quality planning methodologies such as process failure mode and effects analysis (P-FMEA) and control plans by automating key steps, providing AI-driven insights, and streamlining communication among teams. This chapter covers the significance of P-FMEA and control plans, and how AI-powered platforms can optimize these processes for modern manufacturing environments.

P-FMEA: PROCESS FAILURE MODE AND EFFECTS ANALYSIS

P-FMEA is a systematic approach used to identify and evaluate potential failure modes in manufacturing processes, analyze their causes and effects, and prioritize actions to mitigate risks. The goal of P-FMEA is to proactively address potential process failures before they lead to defects, downtime, or safety hazards. Traditionally, P-FMEA involves extensive manual work to collect data, evaluate risks, and determine the appropriate corrective actions. However, AI-based tools, like Kaizen Copilot, can enhance and simplify this process significantly.

In Kaizen Copilot, the P-FMEA module automates much of the data entry, analysis, and risk prioritization, making it easier for teams to collaborate and make informed decisions. The process typically begins by defining the manufacturing process and specifying the key requirements. Users can add detailed information about the process, including team members involved, such as engineers or process leads. The AI helps streamline the creation of FMEA by suggesting potential failure modes, causes, and effects based on historical data or similar processes.

KEY STEPS IN AI-ENHANCED P-FMEA

Defining Process and Requirements: Users define the specific process under evaluation, such as "Manual application of wax inside door panel." The AI can assist by automatically populating common requirements and potential areas of concern based on similar processes.

Identifying Failure Modes: In the failure analysis step, users specify potential failure modes, such as "insufficient wax application." The AI provides classifications for each failure mode – such as controllable, procedural, or noise – and assigns severity ratings (on a scale from 1 to 10), based on the likely impact of the failure. Users can either input data manually or leverage the "AI Suggest" function to auto-generate suggestions for failure modes and their classifications.

CAUSES AND DETECTION

For each failure mode, the AI recommends potential causes and helps users input prevention and detection controls. Users assign detection and occurrence ratings based on how likely the failure is to occur and how easily it can be detected before causing issues. The AI not only suggests these inputs but also automatically calculates a risk priority number (RPN) based on severity, occurrence, and detection values.

RISK MANAGEMENT AND ACTION PLANNING

Once risks have been identified, the system allows users to recommend corrective actions. The AI suggests actions and assigns team members responsible for implementing them. It also provides tools for setting deadlines and tracking the completion of tasks. By automating this workflow, Kaizen Copilot helps teams ensure that actions are taken on time, reducing the chances of recurring failures.

RESULTS AND RISK COMMUNICATION

Upon completion, Kaizen Copilot generates a risk matrix, visually summarizing the distribution of risks (high, medium, low) across the process. The system highlights the top ten most critical failure modes based on their RPN values, helping teams prioritize corrective actions. Additionally, users can update the status of corrective actions and see their impact on the overall risk profile of the process.

CONTROL PLAN

A control plan is a vital document in quality planning, outlining the methods and procedures required to control critical aspects of a manufacturing process. It ensures that all process steps are consistently monitored and that any deviations from quality standards are promptly detected and corrected. The control plan provides a roadmap for maintaining process control, preventing defects, and ensuring that customer requirements are met throughout the production lifecycle.

In the Kaizen Copilot platform, the control plan is tightly integrated with the P-FMEA, enabling users to develop control measures based on identified risks. This seamless integration ensures that the corrective actions and controls recommended during the FMEA process are implemented and monitored in real time.

KEY ELEMENTS OF THE CONTROL PLAN

Process Description

A detailed description of each step in the manufacturing process, specifying the critical parameters that need to be controlled. This may include process temperatures, torque values, material specifications, or other key variables that could affect product quality.

Monitoring and Detection Methods

The control plan outlines the methods and tools used to monitor the process, such as automated sensors, visual inspections, or statistical process control (SPC). Kaizen Copilot's AI enhances this by suggesting suitable detection methods based on the process being evaluated.

Control Limits and Actions

For each process step, the control plan defines acceptable limits (e.g., tolerance ranges) and outlines the actions to be taken if these limits are exceeded. The system ensures that deviations are immediately flagged, and corrective actions are triggered to maintain control.

Roles and Responsibilities

Like in the P-FMEA process, the control plan specifies the personnel responsible for monitoring each process step and executing corrective actions when needed.

AI-DRIVEN ENHANCEMENTS IN QUALITY PLANNING

Kaizen Copilot leverages AI to significantly reduce the time and effort required to conduct P-FMEA and create control plans. The AI's ability to suggest failure modes, recommend preventive actions, and generate risk matrices allows teams to focus more on implementing improvements rather than data entry and manual calculations.

Some of the key AI-driven enhancements include:

AI Suggestion Tools

These tools provide users with automated suggestions for failure modes, causes, controls, and corrective actions based on historical data or similar processes. This helps reduce the risk of human oversight and speeds up the FMEA process.

Risk Prioritization

The AI dynamically updates the RPN values based on severity, occurrence, and detection ratings. It also helps users focus on high-risk areas by highlighting the top failure modes.

Integrated Reporting

The system automatically generates comprehensive reports, including risk matrices and action lists, which can be shared with stakeholders and used for audits or process reviews. Reports can be downloaded in multiple formats, such as PDF and Excel, for easy integration into existing quality management systems.

In summary, AI-based quality planning tools like those offered in Kaizen Copilot enable organizations to perform thorough and efficient risk assessments, mitigate potential process failures, and ensure the consistent quality of their products. By integrating P-FMEA with control plans, manufacturers can proactively address risks and maintain a controlled, high-quality production environment. The automation and AI-driven insights provided by Kaizen Copilot reduce the time required to complete these critical quality planning activities, enabling teams to focus on continuous improvement and ensuring that customer expectations are consistently met.

Case Study: Made-to-Order Seating Facility in Orange County, California
This case study demonstrates the reduction of P-FMEA and Control Plan generation time from several weeks to half an hour for new products.

The challenge was that the made-to-order seating manufacturer was using Excel for P-FMEAs and control plans. Collecting potential failure modes and remedies from all of the stakeholders used to take several weeks for every new process, and didn't benefit from the past P-FMEAs they'd already built.

The immediate impact was that Kaizen Copilot integrated all their existing P-FMEA files, and was able to generate failure modes, potential remedies, and RPN values based on prior data as well as using its own vast dataset, when provided with process steps name. This allowed the manufacturing engineer to arrive at a robust P-FMEA and control plan within half an hour and significantly accelerate the production part approval process (PPAP) process.

In summary, AI is poised to revolutionize process improvement in manufacturing, as demonstrated by the advancements in AI-based tools like Kaizen Copilot. By automating and enhancing key areas such as time studies, line balancing, and quality planning, AI enables engineers to design and implement more efficient, ergonomic, and reliable processes. The insights generated by these AI tools not only improve operational efficiency but also foster continuous improvement across the organization. The case studies presented in this chapter show how manufacturers can reduce labor costs, minimize process defects, and ensure worker well-being through data-driven decision-making. The future of manufacturing process design lies in the integration of AI, allowing for smarter, faster, and more agile process improvement across the industry.

Zeeshan Zia, PhD, is the CEO and cofounder of Retrocausal, a pioneering company in Agentic AI for manufacturing assembly optimization. With a PhD in computer vision and machine learning from ETH Zurich and a master's in electrical engineering from the Technical University of Munich, Dr. Zia brings deep technical expertise to the intersection of AI and Lean operations. His innovations have enabled large-scale deployments with Fortune 500 manufacturers across aerospace, medical devices, and automotive industries. A recipient of multiple prestigious awards – including the ETH Medal, Qualcomm Innovation Fellowship, and Microsoft Best Paper Award – Dr. Zia has also held research and leadership roles at Microsoft, NEC Labs, and Imperial College London. Through Retrocausal, Dr. Zia is driving scalable Industry 4.0 transformations, helping manufacturers reduce errors, optimize workflows, and increase capacity with intelligent automation.

REFERENCES

Egan, M. (2021). American factories are desperate for workers. It's a $1 trillion problem. *CNN Business*. New York.

Faciane, M. (2018). *Reducing the costs of poor quality: A manufacturing case study* (Doctoral dissertation, Walden University).

Maxwell, T. (2023). GitHub copilot, based on OpenAI's codex, helps boost productivity. *Business Insider*.

14 Mixed Model Line Design-AI

Richard Rahn

MIXED MODEL LINE DESIGN OVERVIEW

Lean manufacturing is a big tent. What started out as a unique style of industrial engineering practiced by a few organizations in the 1950s has evolved into a complete House of Lean with a foundation, walls, and roof. Some of the architectural elements that have matured over the years and become a part of the Lean manufacturing business model include Hoshin Kanri, visual factory methods, learning organization goals, SMED, worldwide JIT supply chains, and on and on. The foundation, however, continues to be Lean Industrial Engineering. In today's world of mass customization and the proliferation of different product variants, the term we use is mixed model manufacturing (MMM), which embraces both workflow and material flow design. Without a sturdy foundation in MMM, all the other elements of Lean, as wonderful as they are, will fail to deliver significant performance benefits.

One might surmise that given the importance of MMM, Lean practitioners would be deeply knowledgeable about the MMM method for creating high-performing Lean production lines or value streams. The sorry fact is that most Lean professionals and manufacturing engineers have *not* been trained in this key discipline, and that even after 40+ years of effort in introducing Lean thinking, huge opportunities still exist for learning and improvement.

One way to grow the expertise in MMM is to expand the quantity and quality of training, both in universities and within manufacturing companies. Understanding how to design a mixed model value stream, and how to design its corresponding material flow system, should be considered essential and required Lean knowledge. But if we rely solely on individuals to understand the process deeply and apply it correctly, results will be, shall we say, variable. We also need to provide Lean practitioners with better tools.

The goal of this chapter is to introduce you to the MMM method in more depth and then to explore the opportunities to use AI to improve the line design process. Equally huge opportunities also exist to improve material flow, but that discussion will need to be tackled in a separate chapter.

DEFINITION OF MIXED MODEL LINE DESIGN

Mixed model line design (MMLD) is a manufacturing strategy and methodology that allows the production of multiple product variants on a single assembly line

 DOI: 10.1201/9781003532972-14

without the need for changeovers or significant downtime. This approach is essential in today's market where customization and product model choice are a market requirement. MMLD aims to balance the line to handle various models with different work steps and times, while protecting efficiency, quality, flexibility, and high productivity (Rahn & Leone, 2014b).

The core principles of MMLD involve understanding and analyzing the product mix, determining takt time, balancing the workstations to avoid bottlenecks, and minimizing WIP inventory. The goal is to create a seamless flow of production that can adapt to varying product specifications while maintaining a steady pace and high-quality output.

This strategy is not as exotic or difficult as it might sound. Most manufacturers in today's world are mixed model companies. For many product lines, over the decades the number of different models and variants has grown significantly, while often the old models are not retired. It is not unusual for a company to have experienced a doubling in the number of model variants it needs to produce on the same line during the last ten years. Not only is the work itself much more challenging than before, but the material flow requirements will have expanded at an even greater pace. Getting good at MMM can make the difference between survival and failure, or profitability and loss.

UNDERSTANDING THE MIXED MODEL LINE DESIGN ROADMAP

At an elevated level, the method for designing a mixed model line properly can be organized into nine main phases. Each phase includes numerous tasks that will need to be completed. We call this documentation of the line design method a roadmap. Breaking the process down into nine high-level phases is extremely helpful in organizing our thinking about how to apply AI, enabling us to focus on specific opportunities for improvement without being overwhelmed by too broad a scope. Here is a brief description of each phase:

1. **Assessment and Master Plan**. In this step, we will assess the current state of the production environment and create a deployment plan for the design process.
2. **Analyzing the Flow**. In this step, we will create process flow diagrams (PFDs) for every product to be included in the design. We will also create additional documents like a mixed model process flow diagram and a process flow matrix.
3. **Collecting Process Data**. For every product and process, we need to document the work steps, work times, and a variety of other data elements. We also need to have a forecast of demand or capacity goals for the new design.
4. **Calculating Takt and Resources**. In this step, we will calculate takt time for every process, and then calculate the number of resources (machines and people) in each process needed to meet the target capacity goals.
5. **Designing Workstations and Cells**. In this step, we will assign work steps to individual workstations and attempt to level load the workflow. We also design each workstation for ergonomics and safety.

6. **Brainstorming Product Flows**. In this step, the design team will experiment with different layout options at a block diagram level. This is done prior to deciding on a final design.
7. **Modeling the Flow**. In this step, we will build a discrete event simulation model of the proposal design. The goal is to test the design performance in the face of inevitable changes to mix and volume, with the goal of improving the design and making it more robust.
8. **Formalizing the Design**. In this step, we create a CAD-based detailed layout of the proposed design, including all the necessary details like utility drops, location of aisles and doors, and so on.
9. **Selling the Change**. This ultimate step includes a review of the final design by stakeholders, the creation of a deployment plan, and the approval of necessary funding to make it happen.

For each of these steps, our goal is to provide a list of potential improvements to this design process using current and anticipated AI capabilities.

A HUMANISTIC APPROACH TO AI-DRIVEN IMPROVEMENTS

For each of the nine phases of the MMLD method discussed earlier, there are AI tools that can be applied today (or soon), and there are expected advances in AI that are not yet available but will also be applicable to the phases in time (Aslam, 2023). The role of humans will remain primary and central, and AI will be used as a tool rather than a replacement for the line designer. There are dozens of detailed implementation steps required in a line design project, so while complete automation is eventually possible, human designers will be needed in the near future.

In most cases, the strategies described in this chapter have been created and tested by the author. If not, they are presented with a reasonable expectation that they could be developed, i.e., the tools fall within the existing capability of an LLM. Current AI strategies include the use of AI to write code and macros. Even if the LLM cannot perform data analysis to the level needed internally, it is quite capable of writing an Excel macro to do this. I have used this approach often, especially when I did not have confidence in AI's data processing reliability. Another great current use is to provide AI with detailed instructions of what is desired, a highly specific prompt. This is especially effective if the desired output is text-related, the area where LLMs shine.

We should not overlook the ability of existing LLMs to contribute innovative ideas to the line design project. Presenting a process description and then asking AI to suggest ways to improve will generate a sometimes surprising and unanticipated number of suggestions. This is not necessarily because the AI is being creative, but it does have access to orders of magnitude greater existing information than any individual human and can often identify opportunities that you may not have thought of.

Disclaimer: Given the pace of advancements in AI and given the wide scope of the activities discussed next, it is expected that there will be many AI possibilities already available now that the author is not familiar with. Inevitably, this discussion will be incomplete, so think of it as an appetizer and not a full meal.

PHASE 1: ASSESSMENT AND MASTER PLAN

The goal of this first Phase is to conduct a review of the current state of the target area to be designed and deliver a comprehensive project plan for the design phase of the project. The deliverable at the end of Phase 9 is a dimensionally accurate layout, a detailed text description of the future state value stream, and an implementation plan. The physical implementation phase itself is managed as a separate project, with its own mixed model implementation roadmap.

BUILD A VALUE STREAM MAP

If provided with a sufficiently detailed text description of the value stream, AI can build it, add insights into areas of potential improvement, and calculate manufacturing cycle time (MCT) and value-added performance. AI can also generate the flowchart itself by writing the necessary code, either as Visio or Excel macros or by generating the flowchart directly using Python or other programming language.

For example, a human team could walk the gemba and record a detailed commentary on a production line or value stream. This commentary could include remarks on current performance, accumulation of inventory, equipment reliability, bottleneck resources, estimated cycle times, and so on. The audio recording of this commentary could then be quickly transcribed by an AI tool, and then provided as input to the value stream AI.

A standard script could also be developed to provide detailed instructions to the AI regarding the VSM creation process. This script would be the platform for continuous improvement, adjusting it in response to the AI performance and results.

AI-DRIVEN MASTER PLAN

AI can be provided with a detailed roadmap of the MMLD process, along with a description of each implementation step. Using input from a previously created VSM along with additional data from the implementation team, the AI could build a draft of a complete implementation plan. The quality and usefulness of the plan will be proportionate to the quality of the instructional script, but this is something that can be refined and expanded over time.

AI systems like ChatGPT can also assist with the creation of a master plan using software like Microsoft Project. While it cannot create a project file directly, it can generate a project macro that will accomplish the same thing by executing the macro. The current (August 2024) version of ChatGPT can also generate Excel files directly, without requiring the intermediate use of a macro. These Excel files can be imported directly into the project management software.

ESTIMATING ROI AND QUANTIFYING BENEFITS

An important part of Phase I, especially if significant investment is required, will be the estimate of return on investment. The inputs for this deliverable include financial data like labor costs, inventory dollars, and overtime expense. There is also non-financial information that can be used to estimate benefits, including MCT, market share, and floor space.

Using AI to estimate ROI will require a detailed script that describes each potential source of benefit, what data will be provided, and the method for estimating the future performance of that item. For example, we would expect WIP inventory to be related to reductions in MCT, since if the product can flow through the value stream more quickly it will be in WIP status for less time.

An additional function of an AI-driven ROI analysis is the ability of AI to potentially provide a more in-depth analysis of each benefit source and identify new ones. The AI has in theory been exposed to every book ever written on Lean along with every published case history of Lean improvements!

PHASE 2: ANALYZING THE FLOW

In Phase 2 of the line design roadmap, as a part of the data collection process, the deliverable is to document the products that will be produced on the same production line or value stream. Not every product is a good fit in a mixed model line, and a careful analysis of process commonality, work content variation, physical size, and a variety of other criteria will need to be assessed before finalizing the product family.

Each individual product will need to be documented through the creation of a PFD, showing the required processes and their relationship in time, that is, their work sequence. Individual PFDs are then combined into a mixed model PFD, showing all the possible paths that will be needed in the eventual physical design.

At Phase 2, we will also create the matrix showing the relationship of products and processes called the process flow matrix. This basic matrix format will be used in subsequent phases to document the flow at a deep level, including an assessment of yields, rework, scrap, and work content times for every product/process relationship (Rahn & Leone, 2014a).

Creating Process Flow Diagrams

AI can generate PFDs for each product by parsing process data. A text description of each unique flow for each product can be provided by a designer. A graphical illustration of each PFD can also be produced, with the most reliable method being to ask AI to create an Excel macro.

Mixed Model Process Flow Diagram and Matrix

Once all the PFDs have been provided, AI can integrate individual flow diagrams from the previous step into a mixed model PFD and create a process flow matrix. The key goal here is to be able to visualize the entire product family in both a flow chart and row/column format to finalize the product family determination going forward.

Product Family Analysis

To determine if a product is a good fit in a mixed model family, there are several hurdles that will need to be crossed. The first hurdle is that of *Process Commonality.*

If an individual product does not share a high percentage of the same processes as other products in the family, the resulting process flow will be complex and difficult to manage. AI can perform this analysis and flag the products that fall outside of the recommended guidelines.

Product Demand Analysis

AI algorithms can analyze market trends and historical sales data to generate accurate demand forecasts. The deliverable for this step is to set a target volume value for each individual product, with a special focus on future demand. With access to the history and forecasts for each product, AI can provide a target volume for each product. These values will need to be reviewed by the line designers and approved by human management since they will determine the overall capacity of the line design. Once these target volumes are set, it will become increasingly difficult to change them without impacting the entire design. Providing AI with clear instructions on how to perform the analysis is a great leap forward in generating consistent results for this step.

PHASE 3: COLLECTING PROCESS DATA

Completing Phase 2 in the line design journey gives us a promising idea of the products that we want to mix, the processes that will be required to do so, and the capacity goals we will want to reach. In Phase 3 we will need to dive deeper into data collection with the help of AI. The data collection phase is one of the most time-consuming and labor-intensive, and using AI to reduce the time and improve the accuracy will be hugely beneficial (Beale, 2023).

Sensor-Based Automated Data Collection

AI-powered sensors and IoT devices can collect and document work steps, work times, and other relevant data automatically. These will invariably be specialized use-case tools that are not available in the LLMs like ChatGPT. They can be amazingly powerful and useful but will require additional training and cost.

Sensor-less Data Collection

There are systems currently available that can capture and analyze process-related data without the use of sensors, relying instead on the use of low-cost cameras and vision technology. Several such systems are offered by Retrocausal.ai, their "Kaizen Copilot" and "Assembly Copilot." These again are proprietary systems that use AI but are not a part of the ChatGPT product.

Advances in "Omni" Capabilities

Open AI, the parent company of ChatGPT, is rolling out an advanced version of ChatGPT that includes the abilities beyond text: to communicate verbally in real-time and to process visual information. There are options for using your phone's camera as well. There will be opportunities soon to document work simply by using your cell phone. Imagine, for example, doing a line walk-through while pointing your camera at the line and describing what is happening verbally, and getting back an accurate and useful value stream map or standard work definition!

Checking Standard Work

AI can be extremely helpful in validating, organizing, and optimizing existing standard work documentation. If the work steps, the work times, and other tasks-related information is available in Excel format, AI can read it and provide feedback in a variety of ways: validating the sequence of steps, identifying products that fall outside of recommended time variation guidelines, flagging non-value-adding work steps, implementing check-do-check instructions. With a skillfully constructed prompt, be prepared to be amazed at what can be done!

PHASE 4: CALCULATING TAKT AND RESOURCES

Once we have arrived at Phase 4, the data required for the line design has been gathered and we are ready to perform some simple calculations. The benefits of AI for this Phase are less since doing them in an Excel spreadsheet is straightforward. Since AI has already collected the necessary input data, it still makes sense to build an AI tool to do the work of calculating takt time and resources (number of people and machines) for each process. Using AI will promote consistency and accuracy in the calculations and provide a platform for continuous improvement in this and other line design steps.

Automated Calculations

AI can quickly calculate takt times and resources for each process by drawing on the data already collected in Phases 2 and 3. These calculations are straightforward and can be described to the AI with a clearly written script. The downside of this approach, as discussed previously, is that it will be impossible to directly validate the method that AI has used. If the script is not completely accurate, AI will generate seemingly correct calculations that are in fact wrong. If this is a concern (and it is at this state of AI maturity), consider using the option discussed next.

Excel-Based Calculations

A human touch and quality check can also be achieved by using AI to create a set of macros to perform these calculations instead of doing the calculations directly. In that way, the visual basic for applications (VBA) code can be analyzed directly if there are any questions about the results. AI could provide the core data for each product in a spreadsheet format, and the calculations could be executed by the macros.

Determining the Number of Workstations

Keep in mind that the calculations of takt time and resources are performed at the process level. A single process, final assembly, for example, will most likely require more than one resource or workstation. If this work is to be performed in an assembly line fashion with partially completed work moving from station to station, the AI will need to determine if fractions of a workstation should be rounded up or down. This decision-making step will need to be described clearly in a script.

PHASE 5: DESIGNING WORKSTATIONS AND CELLS

Phase 5 is one of the most important and time-consuming steps. All the calculations done up to now have been at the process level, including the determination of the number of workstations and machines. In this phase, you will need to allocate work steps to individual workstations if the work is to be performed in a sequential way, with each station completing a portion of the total work in the process. This step is commonly called line balancing or level loading and needs to be completed model by model and process by process. While the task of line balancing is not inherently hard, many products and processes can make this one of the more challenging phases to complete due to sheer volume.

Note also that if a change is made to the number of workstations in a process, this line balancing step will need to be re-done to balance the work to the new number of workstations. This is one of the main reasons why a line redesign is so time-consuming, and why help from AI can really speed things up.

Automated Line Balancing

In this approach, the work steps and times for a specific process and model are fed to the AI as text or as an Excel file. Based on clear instructions on how to go about it (the prompt), the AI will allocate tasks to the target number of workstations. It will then return the line balancing plan for that product and process. This will need to be repeated individually for every other product in the product family.

Macro-Based Line Balancing

In this alternative method, the AI will be provided with the same detailed script, but instead of doing the balancing work, it will generate a spreadsheet macro to do the balancing. The advantage of this method is that while it is slightly less automated, it is possible to analyze the line balancing process and adjust if needed by reviewing the VBA code.

Ergonomics Optimization

AI can analyze workstation design for ergonomics and safety, suggesting improvements to reduce strain and injury. It can use drawings and photographs as well as text as input and prompt the user if additional information or detail is needed. The weight of various parts consumed at the workstation can be accessed from the PFEP database and included in the ergonomic assessment.

Material Presentation Design

The AI can access information in the PFEP database and provide a suggested material presentation plan at the workstation level. This will include placing the heavier and more frequently used items closer to the operator, suggesting the necessary shelving and racking, and even recommending smaller container sizes if appropriate.

Advanced Workstation Design

As mentioned earlier, advanced AI systems for workstation analysis are also available from third-party vendors like retrocausal.ai. These camera-based systems can address many of the needs documented above for workstation optimization.

PHASE 6: BRAINSTORMING PRODUCT FLOWS

Now that we have calculated the number of machines and workstations we will need in our line design, and we have gone through the laborious tasks of allocating work steps to individual workstations for each product, we are ready to move to Phase 6, brainstorming product flows. The goal in this step is to experiment with various layout options, and to work through their respective pros and cons, without committing ourselves too early to a final decision. We focus the discussion at the workstation level, but without needing to consider the minute details of the final layout. These will be tackled later in Phase 8.

This phase is normally completed as a team effort, using sticky notes and markers. AI can be useful in providing the team with the most likely options, along with an explanation of the pros and cons of the choices.

AI Layout Optimization

The basic requirements for the layout are already available to the AI and were created in the previous phases. Drawing on this information, AI can generate and evaluate different layout options, helping the design team identify the most efficient configurations.

We can anticipate that in the future, AI will be able to create and validate workflow layouts automatically more completely. For the time being, this phase should be primarily human-based.

PHASE 7: MODELING THE FLOW

The goal of Phase 7 is to build a discrete-event simulation model of the proposed value stream design, based on all the excellent work done up to now. We regard this step as critically important, especially if the project will require a substantial investment of time and money. As we often say, the worst time to be making changes to your design is when the line is already built. Similarly, the worst time to be making changes to your house remodeling project is when you've already built the house. Building a model is a low-cost and powerful way to test the design under varying conditions of product mix and volumes, like testing an aircraft design in a wind tunnel. While simulation modeling is not new, the challenges of learning this software have been an obstacle to widespread adaptation of the method.

There is a plethora of commercial simulation modeling software tools that can be used at Phase 7, and we will also explore the option of building your own with AI. There are two basic approaches we will explore: using AI to build the input information to feed external simulation modeling software and using AI to build the model including the simulation algorithms using Python or another programming language.

It may be necessary after Phase 7 to go back to Phase 5 to perform another line balancing if the results of the simulation experiments indicate that workstations need to be added. If Phase 5 can be largely automated with AI, this will convert a major delay into a minor one.

Using AI to Create the Inputs

All modern simulation modeling software includes some capability to import model data from an external file, often Microsoft Excel. The proposal here is to use AI to create this input data, either in Excel or CSV formats, which can be transferred to the simulation software. The use of this technique assumes that the raw modeling data is available, and that the user is competent in the use of the simulation modeling software (not a given).

As a case study, the author was able to use ChatGPT to write the Excel macros that converted a line design resource worksheet into the seven input worksheets required by the Lean Design Simulator, a proprietary simulation modeling engine designed specifically for value stream optimization. While ChatGPT is not creating the data files directly, the use of intermediate macros provides additional control of the desired output. Using this technique, once the raw data was available, the simulation data file could be created in a few seconds instead of several hours.

Using AI to Write the Code

The proposal here is to use AI to not only format the model data but to also write the simulation modeling code, eliminating the need for commercial simulation software. There are several libraries of simulation model functions available for Python, including the popular SimPy library. AI would utilize this library and generate the code that would then be run in Python. The latest versions of ChatGPT can also execute the code internally, theoretically eliminating the need for external simulation software.

The author has built several proof-of-concept models using this method, and it appears to work. A fully functional model will require a detailed and comprehensive script for the AI to follow, but it could then be reused in the future.

Using AI to Conduct Experiments

Traditionally a team of people will use simulation models to conduct a series of reasonable simulation experiments to test the performance of the system in the face of changes to customer demand and product mix. The experimentation phase can be accelerated significantly with two AI-driven approaches: (1) have AI analyze the process data and provide a set of recommended scenarios to test, based on projects of realistic demand patterns and forecasts, or (2) run these recommended scenarios automatically and provide an in-depth report of findings. Option 2 assumes that the AI has access to the simulation modeling software (or can execute code internally) and can run models independently.

PHASE 8: FORMALIZING THE DESIGN

Phase 8 requires the creation of a detailed and dimensionally accurate layout of the proposed design, bridging the gap between the concept and the eventual physical reality. In architectural terms, this is the creation of final construction documents. The house has not yet been built, but the builders have an accurate set of drawings that they will use. AI can be a useful helper in this process, but there are many details of a physical design that are not available to the AI. For example, the location of utilities like power, water, air, and so on will require the input from facilities and manufacturing engineering and were not a part of the design up to now.

One way to assess the capabilities of AI in a specific use-case is to simply ask. When asked about how it could help, ChatGPT was unusually cautious regarding its ability to help with the creation of a CAD-based layout. For example:

> *I cannot create or manipulate CAD (Computer-Aided Design) drawings. While I can help conceptualize layouts and provide guidance on dimensions and placement, the actual creation of detailed, dimensionally accurate drawings must be done using CAD software like AutoCAD, SolidWorks, or similar.*

While it is easy to imagine a specialized AI system that can automate this step in the roadmap, it does not exist currently (as far as we know). In the future, we can expect important things: assistance in creating detailed CAD layouts, optimization of the placement of utilities, aisles, and doors based on ergonomic and efficiency considerations, generative design techniques, and even AI-powered augmented reality providing immersive visualizations of the proposed design, aiding in stakeholder understanding and approval. For the time being, however, we regard this step as a human-based activity.

PHASE 9: SELLING THE CHANGE

Phase 9, the final phase, captures several elements important in any implementation project: a report on what has been accomplished thus far, and a call to action for future steps. In the case of a factory project, the call to action requires "selling" to management since it will require some investment. In Phase 9, the capabilities of AI can really shine.

AUTOMATED DOCUMENTATION

Since AI now has access to the complete project information, including an upload of the final CAD layout, it can generate detailed reports and presentations to support the final design and deployment plan. Humans can provide a detailed description of the desired report formats to be followed, including section headings and sub-topics. LLMs excel in filling in the details.

REASSESSMENT OF PROJECTED BENEFITS

It is necessary at this phase to reassess the initial benefits calculations projected at Phase 1. The latest performance data can be used, along with refined estimates of the expected performance in productivity, lead-time, quality, inventory, overtime, and other metrics. AI can assist in the development of this new input data, as well as the creation of the benefits report.

IMPLEMENTATION PROJECT PLAN

The AI system as well as the human partners already have experience in the creation of a project plan in Phase 1, and this one will be similar. Instead of using the line design Roadmap as a starting point, AI will use the Implementation Roadmap as a guide. The deliverable will be similar: an electronic project plan that shows the tasks, the task interdependencies, the required resources (people and equipment), the work content times, and estimated costs. Until AI can generate the project file directly, we can interface with software tools like Microsoft Project using macros or importing directly from an Excel workbook that AI will create.

IMPLEMENTATION PRESENTATIONS

AI can generate comprehensive reports and presentations to communicate the final design to stakeholders. These presentations can include interactive elements and simulations to help stakeholders visualize the proposed changes. This falls within the current capabilities of AI tools like ChatGPT.

By integrating these current and future AI capabilities, the MMLD methodology can become more efficient, adaptive, and effective, leading to more robust and optimized production environments. Most of these AI tools assume that a manufacturing engineer is nearby, supplying needed data and validating the AI outputs.

Based on the MMLD steps documented in this chat, what are the possibilities for AI to fully automate the entire line design process? If the demand data and the necessary process data could be provided (by human engineers), could the subsequent resource calculations, workstation optimization, preliminary design, simulation modeling validation, final CAD-level layout, and deployment plan be created by AI with little participation from human designers?

FUTURE POSSIBILITIES FOR AI TO FULLY AUTOMATE THE LINE DESIGN PROCESS

The concept of fully automating the MMLD process from Phase 1 to Phase 9 with AI is ambitious yet increasingly feasible with advancements in AI technology. Below we explore how AI could potentially automate each step of the MMLD methodology, provided that human engineers provide demand data and necessary process data.

The detailed MMLD methodology already exists and would be a part of the AI training content. With the current level of sophistication of the LLM AI systems, an extensive script for each of the MMLD steps would be needed. While this is not a small task, it is certainly achievable.

To achieve full automation of the MMLD process, an integrated AI *system* would be necessary. This system would comprise various AI modules, each specializing in distinct aspects of the design process through a detailed training process (Holweg et al., 2023). These modules would need to communicate and collaborate seamlessly, storing and sharing data and insights to optimize the entire process. Here's a conceptual overview of such an integrated system:

1. **Data Integration Module**: *Gathers and preprocesses demand and process data from various sources, ensuring data quality and consistency.*

   In many companies, this demand and process data already exists in an electronic format, either in an existing ERP/MRP system or in other databases. The steps that a human engineer would go through to gather this data can be scripted and performed by an AI agent.
2. **Assessment and Planning Module:** *Analyzes current production states and generates a master plan using AI-driven project management tools.*

   The detailed MMLD methodology for this step already exists and the AI system could be trained in it. From this, a detailed project plan could be produced with Resource and Time estimates generated for each project deliverable. Existing project management software like Microsoft Project can import AI-generated VBA code to populate a project plan in seconds.
3. **Flow Analysis Module:** *Creates and optimizes process flow diagrams and matrices, ensuring efficient material and product flow.*

   The necessary inputs may be available in existing legacy systems or can be provided by human engineers. The contribution of AI at this step would be not only a clerical one in generating PFDs, mixed model PFDs, and Process Flow Matrices. AI could also provide an initial analysis of the commonality of processes and recommendations for improving the process flow even at this initial stage.
4. **Resource Optimization Module**: *Calculates takt times, determines resource requirements, and assigns tasks to workstations.*

   The basic calculations of takt time and required resources are straightforward, but there are decisions that will need to be made that can be captured in a detailed AI-training script.
5. **Workstation Design Module:** *Designs ergonomic and safe workstations, balancing the workload across the line.*

   The task of "level loading" the work steps across the calculated number of workstations normally will take days or weeks to complete. In a mixed model line, the balancing needs to be done model by model, and AI could reduce the time required and the quality of this work. This step is normally highly time-consuming for human engineers. An ergonomic and safety analysis for each workstation is a task that is often overlooked or not done well, and with proper training, an AI system could improve this requirement.

6. **Layout Generation Module:** *Generates and evaluates multiple layout options, simulating performance to identify the best configuration.*

   Based on inputs from several of the previous AI modules, this step would analyze several options for a "conceptual flow" and provide robust recommendations for the sequence of work across various manufacturing processes. Initial recommendations for reducing MCT and recommended buffers (in-process kanbans) would also be a part of the analysis.
7. **Simulation and Validation Module:** *Builds and tests discrete event simulation models to validate the proposed design.*

   Most simulation modeling software will accept input from external files to build a model of the proposed mixed model value stream. The goal of this step is to run experiments with varying levels of demand and product mix to "stress test" the design. We envision that AI could generate the necessary simulation model files for input into simulation modeling software. AI may even be capable of running the model without needing to use external software systems. It could, for example, code a Python-based simulation model using existing Python simulation libraries and execute the code internally. We can envision the possibility of AI independently running dozens or hundreds of simulation iterations and providing design improvement recommendations based on the simulation results.
8. **CAD and Detailed Design Module:** *Creates detailed CAD layouts, optimizing for various design constraints and requirements.*

   A detailed layout of the proposed new design will be needed before the implementation team can begin the physical transformation work. We can envision giving the AI access to detailed CAD drawings of the current state facility, and having it incorporate the new design automatically.
9. **Change Management Module:** *Prepares presentations and deployment plans, facilitating stakeholder approval and implementation.*

   With a detailed training script, AI could at a minimum produce a comprehensive implementation template to be completed by human engineers and managers.

In the past, these capabilities would fall solidly into the category of science fiction. Within the next few years, we see that as becoming possible.

CONCLUSION

The potential for AI to not only improve but fully automate the MMLD process is vast, with current and anticipated AI capabilities covering every aspect of the methodology. While human engineers would still play a critical role in providing initial data and overseeing the process, AI could handle the bulk of the analytical, design, and optimization tasks. This would lead to more efficient, adaptable, and robust production line designs, enabling manufacturers to meet the evolving demands of the market with greater agility and precision.

Until then many of the opportunities discussed in this chapter are available now, so what is holding us back? Here are some of the hurdles that need to be jumped over to make more complete use of existing LLM tools like ChatGPT and Claude.ai:

1. The engineering and material flow teams need a strong understanding of the line design and material flow roadmaps. Without being subject-matter experts it will be impossible to create the necessary AI prompts, or to validate the output from AI. The team won't know if the AI is hallucinating or not if they don't have the knowledge themselves. Generic Lean training won't be too helpful either; we are talking about Lean Industrial Engineering knowledge.
2. The design team will need some training and (mostly) hands-on experience in writing prompts and interfacing with the AI. Skillfully written prompts are essential in getting the desired outputs from AI. There is even a new rapidly growing career job: Prompt Engineer.
3. If you want to take advantage of the considerable skills of AI in writing macros and code, the team will also need to have some expertise in these areas. It will not be necessary to be a coding guru, but if you can't do simple debugging of an Excel macro, your ability to use AI in this domain will be limited.
4. All these potential AI applications will take time to develop, test, and iterate until you reach the desired output. The organization will need to make this R&D time available to work through these issues. For most of the roadmap steps, there is little capital investment required, but people's time availability is paramount.

Other than these needs, nothing stands in your way! So go out and improve your Lean processes and share your achievements with the world.

SELF-ASSESSMENT: ARE YOU READY FOR MIXED MODEL LINE DESIGN AND AI?

Here are ten questions you can ask yourself to assess if your organization is ready for or in need of an MMLD-AI approach to Lean manufacturing. A high number of Yes answers may point to this need!

1. **Product Variety and Customization**
   - Does your company produce a wide range of different products, or install product options, on a single line and is demand for "mass customization" increasing?

   If yes, the mixed model manufacturing style of Lean could be essential for managing growing complexity while maintaining quality and productivity.
2. **Production Flexibility**
   - How well does your current production system handle changes in product demand, model mix, or production volumes without requiring line changeovers or incurring costly production delays?

Mixed model manufacturing is designed to reduce these disruptions by allowing flexible, continuous production with zero changeover between models.

3. **Bottlenecks and Imbalances**
   - Are bottlenecks or imbalances frequent on your production line, especially when handling a mix of different product variants?

   Mixed model manufacturing aims to balance workstations and workflows to address inefficiencies and keep operations smooth.
4. **Workforce Expertise and Training**
   - Do your teams have the necessary expertise in Lean manufacturing and MMLD? If not, are you prepared to invest in training to build these critical skills?

   Successfully implementing mixed model manufacturing requires foundational knowledge in Lean Industrial Engineering. These skills are arguably the most important.
5. **Data Availability and Management for AI**
   - Does your organization have reliable and detailed data on work content times, process flows, and demand forecasts? These are required for AI analysis and simulation modeling creation.

   AI tools depend on accurate, well-maintained data to optimize processes, allocate resources, and simulate production scenarios.
6. **Process Complexity**
   - How complex are your production processes in terms of product commonality, work content variation, and material flow? Could simplifying this complexity through a structured MMM approach improve operations?

   Mixed model manufacturing helps manage process complexity while maintaining efficiency across product variants like different models and/or options.
7. **Current Use of AI or Advanced Technologies**
   - Is your organization currently using AI or other advanced technologies for tasks like data collection, process optimization, or simulation modeling?

   If your organization is already embracing new technologies, it may be well-positioned to adopt AI tools for mixed model manufacturing.
8. **Cost and ROI Considerations**
   - Are inefficiencies like high WIP inventory, long lead times, or resource imbalances driving up costs? Have you considered using AI to estimate the potential ROI from implementing MMM?

   AI can be a powerful tool to assess current inefficiencies and project the financial benefits of adopting a more flexible production system.
9. **Management and Stakeholder Buy-In**
   - Is there strong support from management for long-term investment in Lean manufacturing practices and AI technology? Are they prepared to dedicate the necessary resources to drive significant changes? If not, this is a clear show-stopper.

Implementing mixed model manufacturing and AI requires leadership commitment, deep knowledge of Lean by the management team, and a willingness to embrace transformation.

10. **Readiness for Future AI Capabilities**
    - Is your organization prepared to leverage future advances in AI, such as automated line balancing, layout optimization, or real-time production monitoring? Do you have the infrastructure in place to support these emerging technologies?

    Forward-thinking organizations with an adaptable infrastructure will be better positioned to take advantage of the next generation of AI-driven manufacturing capabilities.

REFERENCES

Aslam, B. (2023, March 23). Lean methodology and artificial intelligence (AI) to increase productivity. *LinkedIn*. https://www.linkedin.com/pulse/lean-methodology-artificial-intelligence-ai-increase-bilal-aslam/

Beale, S. (2023, July 20). The AI approach to Lean manufacturing. *MasterControl*. https://www.mastercontrol.com/gxp-lifeline/the-ai-approach-to-lean-manufacturing/

Holweg, M., Davenport, T. H., & Snyder, K. (2023, November). How AI fits into Lean Six Sigma. *Harvard Business Review*. https://hbr.org/2023/11/how-ai-fits-into-lean-six-sigma

Rahn, R., & Leone, G. (2014a). *The complete guide to mixed model line design*. Flow Publishing.

Rahn, R., & Leone, G. (2024b). *Material flow*. Flow Publishing.

15 Lean-AI Case Studies

CASE STUDY 1: COMPREHENSIVE SAFETY SOLUTIONS USING LEAN-AI

Safety is one of the primary concerns in most factories. According to the National Safety Council (NSC), the total cost of work injuries in the United States in 2022 was $167.0 billion. In 2022, the cost per worker was $1,040, and the cost per fatality was $1,390,000 (NSC, n.d.). Some international factories are more prone to safety concerns due to a lack of regulations and government oversight, as compared to the United States. This case study highlights how safety in an international factory addresses safety at a holistic level using Lean-AI principles.

Scenario

A US-based organization was operating a manufacturing plant in Latin America. Within the plant, there were significant safety issues like trip hazards, liquid spills, and equipment-related injuries. This resulted in numerous near-misses, accidents, near-death accidents, and even a couple of fatalities. Outside the plant there existed violent gang members that could break in, steal valuables, leave behind dangerous packages, terrorize the factory workers in the parking lots, or even settle grudges with guns. This had a major negative impact on the health, well-being, and absenteeism of the workers, and the morale of the workforce was very low. The executives were aware that to continue operating the plant safely and profitability, immediate action needed to be taken. If the factory could not be secured, the decision would be made to close it permanently and move the operations to a different country.

Executive Decision

The executive decision was to develop a plan to secure the factory from inside and outside using advanced technologies like AI, and the IoT in combination with traditional tools from Lean technology. A team consisting of internal experts and consultants was formed to address this (IBM, n.d.).

Proposed Safety Solutions

The plan was developed for safety solutions for the following:

DOI: 10.1201/9781003532972-15

Tripping Hazards

To safeguard operators in the manufacturing environment from tripping hazards due to obstacles in aisle ways and high traffic areas on the factory floor that result in injuries, recordable incidents, hospitalization, and productivity losses.

Solutions

6S: implement good 6S that makes the factory floor a safe and high-performance workplace.

Perimeter Protection: Using cameras, thermal cameras, and sensor cameras to detect objects thrown or protruding into the target area.

Alerts About Hazardous Objects: AI-mapping of the normal state of the shop floor. AI-based alerts for left behind or protruding objects and detect obstructions in aisle ways as shown in Figure 15.1

Benefits

The benefits of the proposed safety solutions are a reduction in tripping injuries and accidents, improved operator safety, increased productivity, and digital records.

Liquid Hazards

To safeguard operators in the manufacturing environment from falls and slips due to liquids on the floor that result in minor to serious injuries, workday loss, and productivity losses.

Solutions

6S: implement good 6S that makes the factory floor a safe and high-performance workplace.

Leak Detection: A water leak detection sensor designed to detect the presence of any liquid including water, fuel oil, etc.

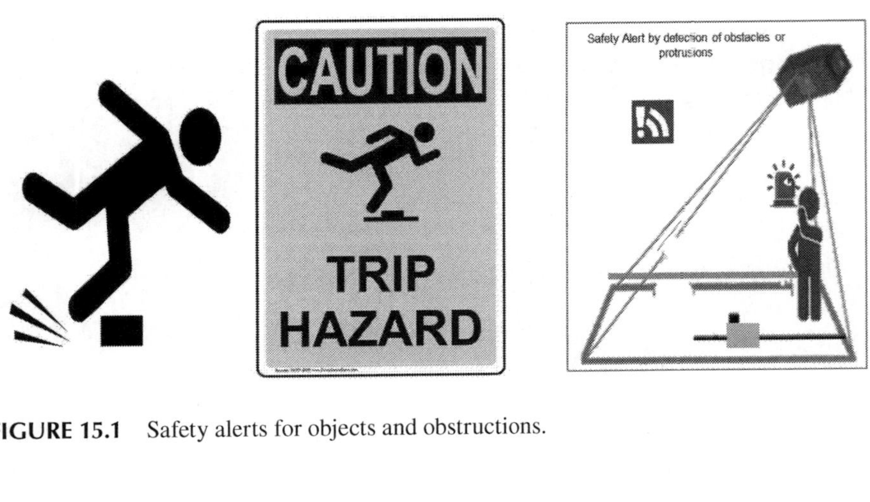

FIGURE 15.1 Safety alerts for objects and obstructions.

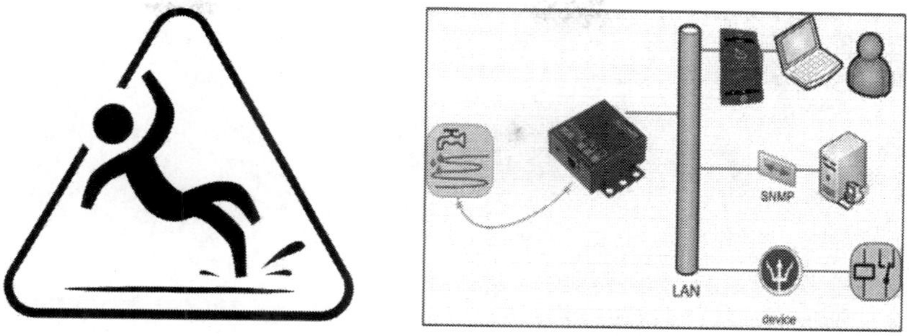

FIGURE 15.2 Safety alerts for liquid leaks.

AI-based Sensors: Floor sensors, wall-mounted sensors, and spot sensors for various applications. The presence of liquids energizes an internal relay to provide real-time visual or audible alert signals to operators, existing alarm systems, building management systems, or SMS messaging systems as shown in Figure 15.2.

Benefits

The primary benefits of the proposed solutions are a reduction of injuries due to slips and falls, improved operator safety, increased productivity, and digital records.

Equipment Hazards

To safeguard against limbs or clothes being caught in restricted equipment areas with moving parts resulting in serious injuries, loss of limbs, hospitalization, workday losses, and productivity losses.

Solutions

Safety measures: Use a 6S map to identify areas of vulnerability in the factory and Implement basic safety measures to prevent break-ins.

Geofencing: An AI-based safety geofencing application that will provide alerts or stop triggers when a worker or others enter an area where there are active moving parts (TechTarget, n.d.)

Cameras and/or 3D LiDAR: Cameras and/or 3D LiDAR (Light Detection and Ranging) will monitor the area for movement and trigger an alert when a person enters an area of potential danger. The trigger will either enable an alarm or provide a hard stop for the machinery (if integration is possible) or an audible alert (Ouster, n.d.) as shown in Figure 15.3.

Benefits

The primary benefits of the proposed solution are the prevention of serious injury due to entering unsafe areas, proactive operator safety, increased productivity, and digital records.

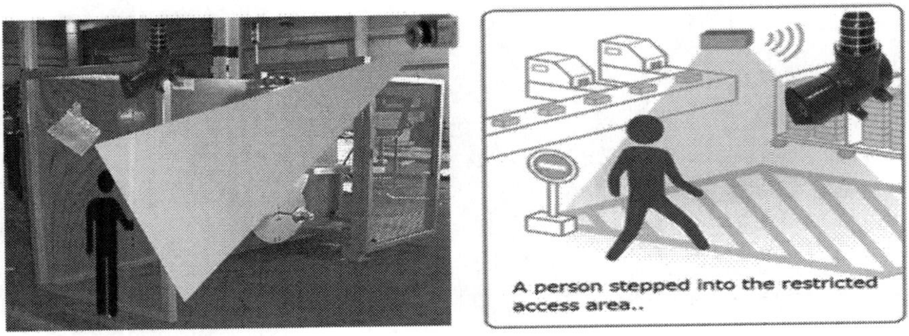

FIGURE 15.3 Safety alerts using cameras/3D LiDAR.

Suspicious Packages Hazards

To secure the factory from left behind or suspicious objects and vehicles by safeguarding the factory perimeter resulting in operator, building, and equipment safety and prevention of injuries, mischief, and sabotage.

Solutions

Visual Controls: Use visual controls (Ortiz & Park, 2018) to ensure that "there is a place for everything, and everything in its place" to ensure visual checks on the factory to prevent any injuries, mischief, and sabotage (Ortiz, 2018).

Perimeter Protection: Use thermal cameras to detect objects thrown into or left at the target area.

Alerts About Hazardous Objects: Provide alerts for left behind or suspicious objects and/or vehicles as seen in Figure 15.4.

Benefits

The benefits of the proposed solutions are increased security, reduced risk of sabotage, damage to factory, equipment and personnel, and digital records.

Intrusion Hazards

To secure the factory from intrusion into the factory area by unauthorized personnel by safeguarding the factory perimeter resulting in operator and equipment safety, and preventing losses due to injury, mischief, and sabotage.

Solutions

Poka-Yoke: Implement poka-yoke to prevent and eliminate safety concerns of workers getting injured near moving parts of equipment

Intrusion detection: Intrusion detection can assist in protecting boundaries and property or improve security if a person or object enters a factory or unauthorized areas.

FIGURE 15.4 Safety alerts using thermal cameras.

FIGURE 15.5 Intrusion detection.

Thermal and night vision cameras: AI-based thermal and night vision cameras to locate unauthorized personnel in restricted areas and alert security personnel as seen in Figure 15.5.

Benefits

The benefits of the proposed solutions are increased security, reduced risk of intrusion and sabotage, operator & equipment safety, and digital records.

Results

The benefits of implementing the Lean-AI solutions were transformative. The trip hazards, liquid spill hazards, near-misses, and accidents were reduced by almost 90%. The plant was secured from the outside bad elements. Break-ins and suspicious package incidents were eliminated. The parking lot was secure. Health, well-being, and absenteeism of the workers and the morale of the workforce were significantly increased. Internal and external customer satisfaction was improved resulting in increased profitability. The decision was made to continue operating the factory with an annual review and upgrade of the AI technology.

CASE STUDY 2: KNOWLEDGE TRANSFER BY ADVANCED TRAINING AND COACHING SOLUTIONS USING LEAN-AI/AR/VR

SCENARIO

A global oil and gas corporation operated a complex facility with around 300 technical employees. New employee training included 12 months of rotation through various areas and the requirement for new employees to shadow "expert techs" for three to four years. All "expert techs" needed to be board-certified, which took ten years. Only about 30% of the current employees were board-certified. The cross-training of employees was very little to none. The company faced significant challenges, including a high percentage of employees with less than three years of experience, complex and lengthy standard operating procedures (SOPs), and the impending retirement of long-term expert technicians. The facility operated 24/7, with shifts of five to ten people, adding to the complexity of operations.

CHALLENGES IDENTIFIED

During a meeting with the executives, several critical challenges were identified as seen in Figure 15.6:

Employee Training: New employees undergo a six-month rotation but take up to ten years to become board-certified, primarily due to the lack of cross-training and reliance on shadowing expert technicians for several years.

Demographic Shifts: With 10–15 expert technicians retiring annually, there was a risk of significant knowledge loss. Additionally, new employees were leaving after two to three years, further exacerbating the experience gap.

Technological and Cultural Changes: The company was experiencing technological advancements and cultural shifts, with employees no longer willing to follow procedures without understanding the reasoning behind them.

IDEAL REQUIREMENTS

The primary objective of the organization was rapid knowledge transfer using advanced technologies. Training time was not a significant concern. Instead, the focus was on improving employee capabilities, speeding up board certification,

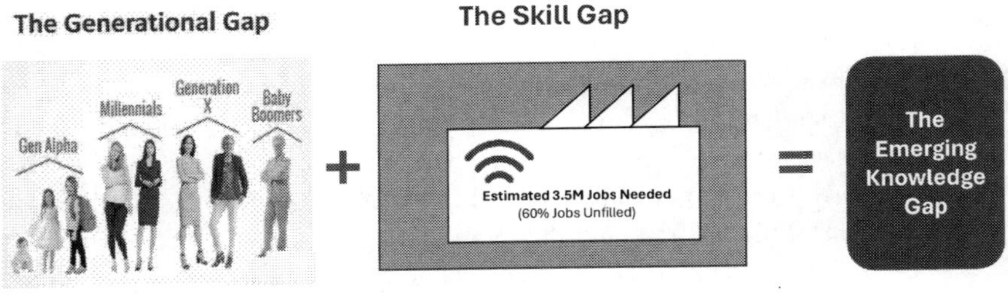

FIGURE 15.6 Organizational challenges.

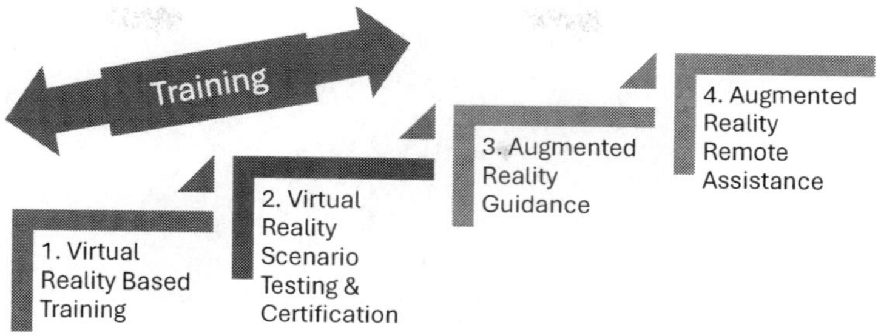

FIGURE 15.7 AR/VR-based solution.

reducing overtime, and enhancing troubleshooting using Lean poka-yoke tools and root cause analysis (RCA) skills. The proposed solution needed to address these key areas while considering the operational risks and the complexity of the SOPs.

Solution

A pilot area was selected for a comprehensive training and knowledge transfer program grounded in the principles of Lean-AI and Industry 4.0, leveraging emerging technologies like VR and AR. The solution was designed to be immersive, engaging, and aligned with the scientific principles of adult learning as seen in Figure 15.7.

Key Components of the Proposed Solution

Pilot Proof of Concept (PPoC): The PPoC involved a detailed analysis of existing training methods, the design of VR-based training and certification, and the implementation of a pilot program. The effectiveness of the training was to be evaluated, with adjustments made based on feedback. Following the successful PoC, the program would be rolled out plant-wide, considering safety, operational, and financial impacts, and ensuring a positive return on investment (ROI).

Implement a Comprehensive Training Model: The decision was made to develop the training program using the ADDIE framework, as shown in Figure 15.8.

Analyze: Reviewed incidents, SOPs, and conducted interviews with certified technicians to understand the current training gaps.

Design: Developed VR and AR-based training solutions, selecting the most effective technology and training partners.

Develop: Created VR and AR solutions, ensuring they address the specific needs of the plant and its workforce.

Implement: Piloted the solutions in selected areas, trained the technicians, and established a support strategy.

Evaluate: Using Kirkpatrick's Four-Level Training Evaluation Model, assessed the effectiveness of the training in the pilot area, adjusting as necessary (Academy to Innovate HR, 2023).

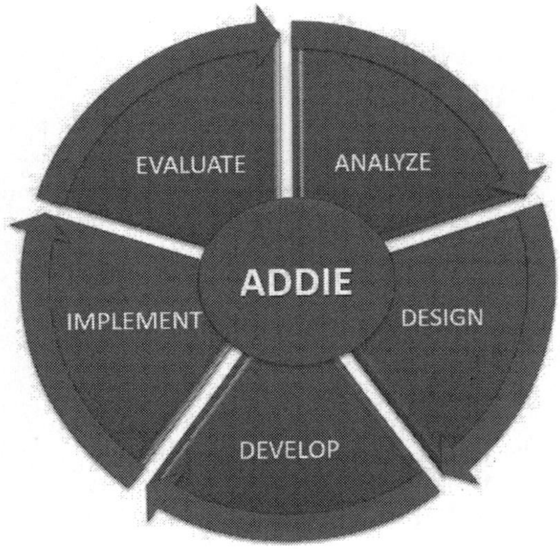

FIGURE 15.8 ADDIE model.

Training Approach

Use VR- and AR-Based Solutions

The key differences between VR and AR are:

Devices: VR requires a headset, whereas AR can be accessed through devices like smartphones.
Environment: AR uses the real world as its setting, whereas VR is entirely virtual.
User Control: AR allows users to stay aware of and control their presence in the real world, whereas VR immerses users completely, controlled by the system.
Experience: AR enhances both the real and virtual worlds, whereas VR enhances only a fictional reality (Tulane University School of Professional Advancement, 2023).

VR-Based Training

This method provided immersive and realistic scenarios in a safe and controlled environment, improving retention and recall. VR training was conducted in a classroom setting, saving time and resources.

Advantages of VR-Based Solutions

The VR-based solutions have significant advantages as seen in Figure 15.9:

Immersive and realistic.
Little/no risk as used in a comfortable and safe controlled area.

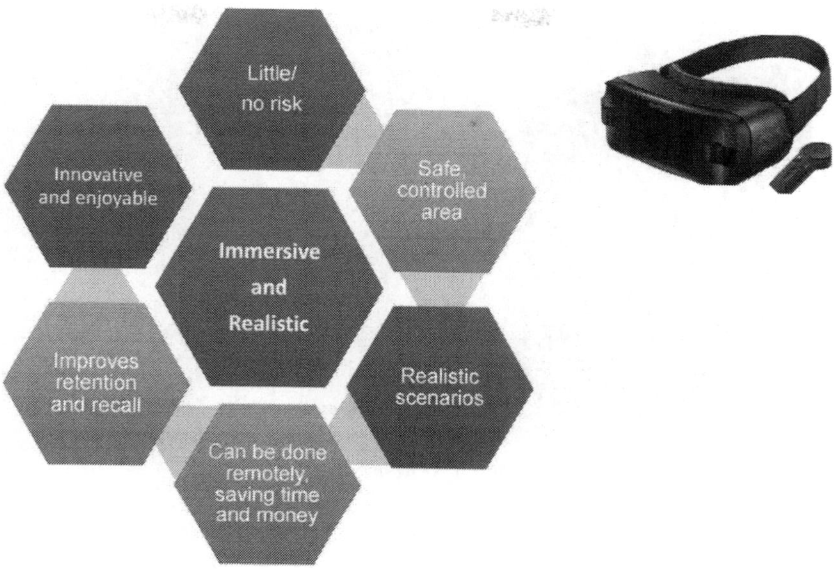

FIGURE 15.9 Advantages of VR-based solution.

Realistic scenarios created using AI.
The gaming elements make it fun and increase retention and recall.
Can be used for certification by creating problem-solving scenarios using AI.

AR-Based Training

AR offered remote assistance, real-time coaching, and guidance through SOPs and labels overlaid on a user's view, making training individualized and context-specific, as shown in Figure 15.10.

Advantages of AR-Based Solutions

The AR-based solutions have significant advantages
The worker is in the real environment having access to virtual data.
Facilitates remote assistance by supervisors from a central location or control center.
Supervisors can provide assistance via documents, videos, and other means.
The settings can be customized to facilitate an optimal coaching experience.

Results

This solution represented a forward-thinking approach to tackle the challenges of knowledge transfer in a rapidly evolving industry. Integrating VR and AR into the training process, aimed to not only preserve but also enhance its operational efficiency and workforce competence, ensuring long-term success in a competitive market.

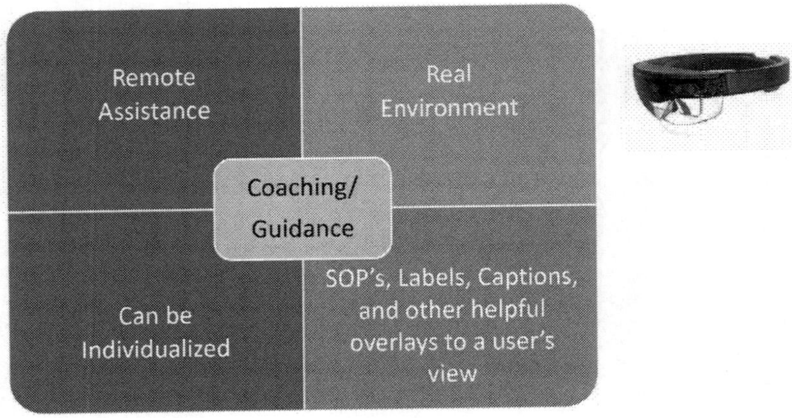

FIGURE 15.10 Advantages of AR-based solution.

The pilot comprehensive training program leveraging cutting-edge technology, was successful in closing some of the knowledge transfer gap and somewhat reduced the risk associated with the retirement of expert technicians and improving the overall capability of the workforce. A longer-term study is needed to understand the full impact on the culture and transformation of the organization.

CASE STUDY 3: LEVERAGING ARTIFICIAL INTELLIGENCE TO ENHANCE LEAN MANUFACTURING

Company Background

This case study organization, founded in 1995, specializes in the production of high-precision electronic components used in telecommunications and consumer electronics. The company operates three manufacturing facilities across North America and employs approximately 3,000 workers. Known for its innovation in electronics manufacturing, the organization has consistently invested in modernizing its operations. Despite its commitment to Lean manufacturing principles, the organization faced challenges related to production inefficiencies, equipment downtime, and inconsistent product quality. In response, leadership sought to integrate AI into its Lean strategy to enhance decision-making, reduce waste, and foster a more innovative culture.

Prior to integrating AI, the organization adhered to traditional Lean principles such as VSM, 6S methodology, and kaizen events. While these tools improved operational processes, the company's leadership recognized their limitations in addressing dynamic market demands and rapid technological advancements (BCG Henderson Institute, 2021). A shift toward incorporating AI-driven tools marked a transformative step in the organization's journey to becoming a model of digital Lean manufacturing.

Application of Lean Tools Enhanced by AI

Value Stream Mapping

VSM is a fundamental Lean tool used to visualize material and information flows within a production system (Rother & Shook, 1999). Traditionally, VSM relied on manual data collection and analysis, which could be time-consuming and prone to errors. The organization revolutionized its VSM approach by integrating AI-powered analytics platforms that continuously monitored production lines. Sensors installed across the production floor captured real-time data, which was fed into an AI-driven analytics engine. The AI identified bottlenecks, non-value-added steps, and areas of delay (Davenport & Ronanki, 2018).

Outcome: This dynamic approach allowed for the creation of live, adaptive VSM charts that updated automatically, enabling real-time decision-making. Lead times decreased by 18% as processes were optimized based on AI-generated insights. Additionally, the organization used AI simulations to test various workflow scenarios, which minimized disruption during process redesigns (Rother & Shook, 1999).

Predictive Maintenance and Overall Equipment Effectiveness

Equipment downtime posed a significant challenge for the organization. The company deployed predictive maintenance systems that utilized AI to analyze historical machine data and predict failures before they occurred. The organization equipped its machinery with IoT-enabled sensors to monitor variables such as temperature, vibration, and operating cycles. AI algorithms analyzed this data to identify patterns indicative of potential failures (Harvard Business Review, 2022).

Outcome: Maintenance schedules became proactive rather than reactive, improving OEE by 22% and significantly reducing unplanned downtime. Maintenance costs decreased as repairs were conducted only when needed, based on AI forecasts. Additionally, the use of digital twins – virtual models of physical systems – allowed teams to simulate maintenance scenarios, further reducing risks and inefficiencies.

Automated Quality Control through Machine Vision

Traditional quality control methods relied heavily on human inspectors, which led to variability in defect detection (Womack & Jones, 1996). The organization addressed this by integrating AI-powered machine vision systems. High-resolution cameras and AI models trained on vast datasets of defective and defect-free products were installed at quality checkpoints. The system was programmed to detect anomalies such as cracks, misalignments, or surface imperfections.

Outcome: Defect rates dropped by 30% due to consistent and precise inspections. The automated system provided feedback to production teams in real time, enabling immediate corrective actions. Over time, the AI system adapted to detect emerging defects, ensuring sustained quality improvements.

AI-Driven Demand Forecasting and Inventory Management

Lean manufacturing emphasizes JIT inventory management to minimize excess stock and waste. The organization integrated AI-driven demand forecasting to streamline inventory processes. AI models analyzed historical sales data, customer trends, market conditions, and supply chain variables. These insights allowed for accurate demand predictions and inventory optimization (Gartner, 2021).

Outcome: Inventory costs were reduced by 15%, while order fulfillment rates improved. AI also flagged potential supply chain disruptions, enabling the company to adjust procurement strategies proactively. The implementation of AI-enabled warehouse automation further optimized storage and retrieval processes, reducing labor costs and inventory errors.

6S Methodology

The 6S methodology focuses on workplace organization through the principles of safety, sort, set in order, shine, standardize, and sustain. The organization used AI to enhance each of these steps:

Safety: Based on thermal mapping and LiDAR technology data, AI identified several areas of safety concern in the shopfloor area, walkways, and also with forklift and other heavy equipment traffic.

Sort: AI identified underutilized tools and materials by analyzing equipment usage patterns, facilitating their removal from the workspace.

Set in Order: AI-designed simulations optimized the placement of tools and equipment for efficient workflow.

Shine: Predictive cleaning schedules based on AI-driven analysis ensured workspaces remained clean and safe.

Standardize: Digital dashboards monitored adherence to 6S principles, providing real-time feedback.

Sustain: Gamified apps encouraged employee participation in maintaining 5S standards, fostering a culture of accountability.

Outcome: These AI enhancements ensured that the 6S principles were consistently applied across all facilities, leading to higher operational efficiency and employee satisfaction.

Challenges and Solutions

Employee Resistance to AI Integration

Initially, workers feared job displacement due to automation. The organization addressed this by emphasizing AI as a tool to assist rather than replace employees. The company launched training programs to upskill workers, focusing on AI literacy and process improvement (McKinsey & Company, 2020).

Solution: AI training sessions were tailored to different employee levels, using hands-on demonstrations to show how AI enhanced their roles rather than eliminating them. The organization also created a mentorship program where early adopters of AI tools guided their peers.

Data Integration and System Compatibility

The integration of AI required consolidating data from various sources. The organization partnered with technology vendors to implement a robust data infrastructure that seamlessly connected AI systems with existing ERP and MES platforms.

Solution: Dedicated IT teams ensured smooth data migration and integration. A phased implementation approach minimized disruptions to operations.

Cost of AI Implementation

Deploying AI solutions required significant investment. The organization justified expenditures through a phased implementation strategy, demonstrating quick wins such as reduced downtime and improved yield rates to gain stakeholder support.

Solution: ROI was tracked meticulously for each phase, and early successes were used to secure further investment. The organization also leveraged government grants and industry partnerships to offset costs.

Impact on Organizational Culture

AI-enhanced Lean practices led to a profound cultural transformation throughout the organization. Employees became more engaged as they participated in data-driven problem-solving initiatives. The company adopted a continuous improvement mindset, fostering collaboration between engineering, IT, and production teams.

Upskilling and Empowerment: Employees developed new competencies in AI tools, enhancing their confidence and job satisfaction. Training programs were designed to make AI accessible, equipping workers with skills to interpret and act on AI insights.

Collaborative Innovation: AI-enabled transparency encouraged cross-department collaboration. Regular innovation workshops brought together diverse teams to brainstorm AI applications in Lean practices.

Recognition and Rewards: The leadership recognized employees who effectively used AI to enhance Lean operations, fostering a sense of ownership and pride. This included incentive programs for innovative AI-driven solutions.

Shift in Leadership Practices: Managers transitioned from traditional decision-making to data-driven strategies, creating a more agile and responsive leadership style. Leaders encouraged experimentation with AI, empowering teams to take calculated risks.

Continuous Improvement Culture: AI-driven insights fostered a bottom-up approach to innovation. Employees suggested enhancements to AI systems, integrating them into Lean methodologies for improved outcomes. This grassroots engagement solidified a culture of continuous improvement.

Results and Measurable Outcomes

Reduction in lead time: 18%
Improvement in OEE: 22%
Reduction in defect rates: 30%
Decrease in inventory costs: 15%
Increase in employee engagement scores: 100%
Increase in AI literacy among employees: 200%
Increase in cross-department collaboration metrics: 150%

Conclusion

This organization's case exemplifies how AI can enhance Lean manufacturing to drive efficiency, quality, and cultural transformation. The integration of predictive analytics, machine vision, and AI-driven demand forecasting enabled the organization to overcome traditional Lean implementation barriers. As AI continues to evolve, companies that embrace its potential in Lean strategies will be better positioned to achieve sustainable competitive advantage.

Moreover, the cultural benefits of AI adoption in the organization illustrate that digital transformation is not solely a technological challenge but also an opportunity to create a more engaged, knowledgeable, and forward-thinking workforce. By integrating AI into Lean practices while fostering a supportive learning environment, organizations can not only optimize processes but also empower employees to become active contributors to continuous improvement.

REFERENCES

Academy to Innovate HR. (2023, June 29). *The ADDIE model: A complete guide to the ADDIE model.* https://www.aihr.com/blog/addie-model/

BCG Henderson Institute. (2021). *AI and lean: Building the factory of the future.* https://www.bcg.com

Davenport, T. H., & Ronanki, R. (2018). Artificial intelligence for the real world. *Harvard Business Review.*

Gartner. (2021). *Emerging AI trends in manufacturing and supply chain.* https://www.gartner.com

Harvard Business Review. (2022). *AI-driven operational improvements in manufacturing.* https://hbr.org

IBM. (n.d.). *What is the internet of things (IoT)?* https://www.ibm.com/topics/internet-of-things

McKinsey & Company. (2020). *The AI advantage in manufacturing operations.* https://www.mckinsey.com

National Safety Council. (n.d.). Work injury costs. *Injury Facts.* https://injuryfacts.nsc.org/work/costs/work-injury-costs/#:~:text=The%20total%20cost%20of%20work,administrative%20expenses%20of%20%2454.4%20billion

Ortiz, C. A. (2018). *Visual controls: Applying visual management to the factory.* Routledge. https://www.routledge.com/Visual-Controls-Applying-Visual-Management-to-the-Factory/Ortiz/p/book/9781138464186

Ortiz, C. A., & Park, M. (2018). *Visual management for lean teams.* CRC Press.

Ouster. (n.d.). *What is lidar?* https://ouster.com/insights/what-is-lidar

Rother, M., & Shook, J. (1999). *Learning to see: Value stream mapping to add value and eliminate Muda.* Lean Enterprise Institute.

TechTarget. (n.d.). Geofencing. *TechTarget.* https://www.techtarget.com/whatis/definition/geofencing

Tulane University School of Professional Advancement. (2023, April 17). *What's the difference between AR and VR?* https://sopa.tulane.edu/blog/whats-difference-between-ar-and-vr

Womack, J. P., & Jones, D. T. (1996). *Lean thinking: Banish waste and create wealth in your corporation.* Simon & Schuster.

16 Guidelines to Lean Practitioners to Adapt to the Future of Lean 5.0+

The Evolution of Industrial Revolutions

It is important to understand the evolution of the industry from 1.0 to 4.0 (Institute of Entrepreneurship Development, 2021) as seen in Figure 16.1:

INDUSTRY 1.0

The first industrial revolution dates back to the late 18th century when mechanical production facilities and steam-powered machines were introduced to help workers in manufacturing. This increased efficiencies and helped small businesses grow into large organizations. Industry 1.0 laid the foundation of modern industries.

INDUSTRY 2.0

Industry 2.0 began at the start of the 20th century with the development of machines running on electrical energy. Electrical machines were more efficient and cost-effective to operate and maintain than their steam-powered counterparts, leading to more productive factories. The first assembly line was built by Ford Motor Co., during the Second Industrial Revolution, leading to streamlining mass production for numerous industries.

INDUSTRY 3.0

Industry 3.0 began in the last few decades of the 20th century, as a result of the proliferation of the electronics industry. Automation of machinery due to the invention of transistors and integrated circuits resulted in less operator effort, higher equipment speeds, accuracy levels, and freeing up the workforce for more creative tasks.

This also spurred the requirements for software systems and a need for management processes like enterprise resource planning, inventory management, shipping logistics, scheduling, and supply chain management. This was the beginning of the modern IT industry.

DOI: 10.1201/9781003532972-16

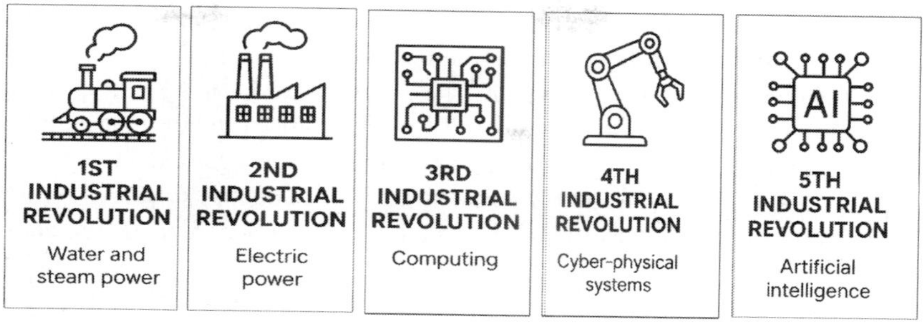

FIGURE 16.1 The Industrial Revolutions.

INDUSTRY 4.0

Industry 4.0 started around 2010 due to the rapid technological shifts that represented a significant advancement in the industrial revolution. Industry 4.0 is built on nine technology innovations and a combination of these technologies results in smart, intelligent, and autonomous systems. These technologies include big data, AI analytics, blockchain technology, autonomous vehicles and drones, nanotechnology, cloud computing, augmented reality, the Industrial Internet of Things (IIOT), 3D printing, autonomous robots, digital twins, and cybersecurity (Schwab, 2017).

These technologies have erased the boundaries of the physical and virtual world, and cyber-physical systems (CPS) further blurred this boundary in manufacturing, introducing numerous other new technologies into the industry. CPS facilitates sharing, analyzing, and guiding intelligent actions using real-time data for different manufacturing processes creating a "Smart Factory." This smart equipment monitors, detects, and predicts failures, PM measures, and remedial action before any downtime. The fourth Industrial Revolution is more than just an extension of the third revolution; it is qualitatively and quantitatively different, marked by a fusion of technologies that blur the lines between the physical, digital, and organic domains.

From factories fueled by the steam engine to the world wide web, big data, and "Smart Factories," each industrial revolution has been a catalyst for bringing a transformative change in our lives.

THE FUTURE OF INDUSTRIAL REVOLUTIONS

Industry 5.0 and Beyond

The rate of change of the pace of the Industrial Revolution is exponential with the innovation of AI itself driving innovation, where AI development propels further AI development, resulting in an unprecedented rate of growth. The change that took centuries, is now happening in decades. The fourth industrial revolution is slowly transitioning into the Fifth Industrial Revolution (Krüger, 2023).

The next revolution will be driven by the rapid evolution of AI, like generative AI, GPT models, and several unknowns. AI models will be more than mere tools or techniques; they will be cognitive partners to human beings, redefining societal structures, global economies, and the fundamentals of human existence. The Fifth Industrial Revolution will usher in an era where AI is an essential, and core component of our thinking processes, influencing our decision-making, creativity, and day-to-day interactions with the real world. AI in conjunction with robotics, autonomous vehicles, and other powerful technologies, AI will transform this planet in ways that are difficult to predict but will be nothing short of being truly futuristic and exponentially revolutionary.

Lean 5.0

Lean 5.0 is the future of Lean technology when Lean-AI will become the new standard of Lean. Lean 5.0 refers to applying Lean principles and methodologies within the context of the Fifth Industrial Revolution (Industry 5.0) and beyond. It builds on the foundation of Lean thinking, which emphasizes eliminating waste, optimizing processes, and creating value for the customer, and adapts these principles to a future where human-centric innovation and collaboration with advanced technologies like AI, robotics, and automation are the new core elements of Lean (Nahavandi, 2019).

HOW AI IS RESHAPING THE CORPORATE CULTURE

AI is fundamentally reshaping corporate culture (Dalal, 2024a), presenting both transformative opportunities and significant challenges. As organizations increasingly adopt AI technologies, the nature of work, leadership, and team dynamics is evolving dramatically. This shift not only automates routine tasks but also enhances decision-making processes, fostering a culture of innovation.

One of the most prominent impacts of AI on corporate culture is its ability to streamline operations through automation. By handling repetitive and time-consuming tasks, AI allows employees to focus on more complex, creative endeavors. This shift is fostering a new corporate ethos where skills like critical thinking, emotional intelligence, and creativity are increasingly valued. For instance, AI-driven tools can automate data entry and reporting, freeing up time for employees to engage in strategic problem-solving, thereby nurturing a culture of innovation.

In addition to operational efficiency, AI is transforming the decision-making landscape within organizations. AI-powered analytics provide data-driven insights that support more informed and objective decisions. As a result, there is a growing expectation among employees to back up their proposals with solid data, leading to a culture of evidence-based management. This transition encourages a more analytical approach to strategic planning, which is vital for organizations striving to stay competitive in an increasingly data-centric business environment.

Furthermore, AI is revolutionizing customer interactions, enabling companies to offer personalized experiences at scale. AI systems handle routine inquiries, allowing human employees to manage more complex, emotionally nuanced customer interactions. This evolution is leading to a corporate culture that values both efficiency and high-touch service, enhancing customer loyalty and trust. AI can also predict potential issues by monitoring customer behavior, fostering a proactive problem-solving culture.

However, the integration of AI into corporate settings is not without its challenges. One significant concern is the potential for job displacement. As AI takes over routine tasks, employees may face uncertainty about their job security, leading to a culture of fear and disengagement. To mitigate these risks, organizations must invest in upskilling their workforce, ensuring that employees can transition into new roles that leverage human skills, creativity, and emotional intelligence – qualities that AI lacks.

Additionally, the over-reliance on AI can result in a depersonalized work environment. As companies become increasingly dependent on algorithms, there is a risk of eroding essential human elements like empathy and intuitive decision-making, which are crucial for maintaining team cohesion and morale. Leaders must be vigilant in promoting meaningful human interactions to counterbalance the technological shift.

Ethical considerations also loom large in the AI landscape. The pervasive use of AI raises concerns about privacy and surveillance, impacting trust and transparency within organizations. Employees and customers alike may question how their data is being utilized, leading to resistance against AI adoption. Therefore, corporate leaders must establish policies that prioritize ethical AI practices and transparency, ensuring that the deployment of AI technologies aligns with the values of the organization.

The integration of AI into corporate culture is a multifaceted endeavor that requires careful navigation. Leaders must embrace the transformative potential of AI while remaining attuned to its challenges. By fostering a culture that values human creativity and emotional intelligence alongside technological advancements, organizations can thrive in the AI era. The successful companies of the future will be those that not only harness the power of AI but also prioritize the human elements that drive organizational success.

ORGANIZATIONAL CULTURE AND LEAN 5.0

Although Lean 5.0 will usher in a seismic shift in the application of Lean technology, the core principles of Lean that must endure this transition are the cultural pillars and humanistic elements of respect and humility. Although AI will significantly enhance Lean in all areas, it can never replace the importance of humanity.

Since the human side of Lean technology stands on these two pillars of respect and humility, without them Lean 5.0 would lack the focus on the human and cultural elements, leaving it just a set of advanced tools and techniques versus a living and evolving philosophy.

Some other ideal behaviors of great organizational cultures like trust, collaboration, innovation, and empowerment must not be sacrificed by AI.

Key Features of Lean 5.0

Human-Centric Focus: Unlike Industry 4.0, which focuses on the digitalization of industries, Industry 5.0 emphasizes the collaboration between humans and machines. Lean 5.0 focuses on leveraging technology not to replace human labor but to enhance human capabilities and creativity.

Integration of AI and Automation: Lean 5.0 integrates AI, robotics, cognitive computing, and other advanced technologies into traditional Lean processes to improve decision-making, streamline workflows, and optimize performance.

Mass-Customization: In Lean 5.0, processes are designed to be highly adaptive to individual customer needs, aided by AI's ability to handle complex data and predictive analytics. This allows for greater personalization of products and services.

Augmented Workforce: By utilizing cobots and AI-driven tools, the workforce in Lean 5.0 can focus on higher-level, creative, and strategic tasks, while automation handles repetitive or hazardous work, boosting productivity, efficiency, and job satisfaction.

Sustainability and Ethics: Lean 5.0 aligns with the broader goals of Industry 5.0, which include sustainability, ethics, and creating value for society. This involves using Lean principles to reduce resource consumption, enhance environmental stewardship, and develop more ethical practices in technology-driven environments.

In essence, Lean 5.0 evolves Lean management by embedding it within the frameworks of Industry 5.0, aiming to harmonize the capabilities of machines and humans to drive innovation, efficiency, and ethical advancements.

Lean 5.0+ will usher in unique opportunities and challenges for Lean practitioners and Lean Leaders. They must adopt a new way of thinking, planning, and executing Lean technologies across organizations. Below are some guidelines for Lean practitioners and Lean Leaders to optimize their effectiveness and efficiency in the era of Lean 5.0+

SOME GUIDELINES TO LEAN PRACTITIONERS TO ADAPT TO THE FUTURE OF LEAN 5.0+

Embrace Human-Machine Collaboration

In Lean 5.0+, practitioners should focus on integrating human intelligence with AI-driven automation. Recognize the value of human creativity, judgment, and emotional intelligence, while leveraging technologies like AI, robotics, and IoT to optimize operations.

Foster a Human-Centric Approach

Lean 5.0+ prioritizes the human element. Encourage a workplace culture that emphasizes employee empowerment, well-being, and personal growth. Focus on creating systems that enhance human work, rather than replace it, aligning with the ethical and sustainability goals of Industry 5.0.

Integrate AI and Cognitive Technologies

Lean practitioners must harness AI to optimize decision-making, predict trends, and enhance processes. By adopting data-driven approaches and cognitive computing, organizations can boost productivity, reduce waste, and make more informed decisions in real time.

Adapt to Mass-Customization

Industry 5.0 emphasizes personalization. Lean practitioners should implement flexible production systems that cater to individual customer needs. This requires integrating advanced technologies to offer mass customization while maintaining Lean efficiency. Mass customization is a manufacturing technique that allows businesses to produce custom products for large markets while maintaining the low cost of mass production (Pine, Peppers, & Rogers, 1997).

Focus on Sustainability and Ethical Practices

As Lean 5.0+ evolves, sustainability becomes central. Ensure that Lean processes are designed to reduce environmental impact, waste, and energy consumption. Additionally, integrate ethical considerations into AI implementation, fostering socially responsible practices.

Leverage Continuous Learning and Upskilling

As new technologies emerge, encourage a mindset of continuous learning. Lean practitioners should invest in upskilling the workforce to ensure they are proficient in using AI and other emerging tools, fostering adaptability to future changes.

Prioritize Data-Driven Lean

Use real-time data analytics and advanced machine learning to drive Lean improvements. Data should be continuously collected and analyzed to enhance process optimization, allowing for more precise and proactive management of operations.

By adhering to these guidelines, Lean practitioners will be well-prepared to navigate the evolving landscape of Lean 5.0+, where human creativity, AI, and sustainability converge.

GUIDELINES FOR LEADERS IN LEAN ORGANIZATIONS TO ADAPT TO LEAN 5.0+

Along with Lean practitioners, the leaders of organizations on the journey to Lean transformation must adapt to the evolving landscape of Lean 5.0+, where human creativity, AI, and sustainability converge.

In the age of AI, leadership can take three paths (Dalal, 2024b):

Technology-Driven Leaders (TDL)
Human-Centered Leaders (HCL)
Tech-Savvy Humanistic Leaders (TSHL)

The TDL will prioritize AI over the fundamental rights of humans. When taken to an extreme with autonomous weapons and allowing AI to make life-or-death decisions, could threaten the very existence of humanity. HCL will prioritize steering humanity toward greatness but without the use of AI technology. This could prevent humanity from reaching its true potential in the fields of science, medicine, space, and other endeavors. TSHL are ideal for the age of AI, as they use empathy with an insight into technology, demonstrate visionary thinking with ethical considerations, and use both human and AI to allow humanity to survive and thrive.

Some Guidelines for developing TSHL are:

Emphasize Ethical Leadership: As AI becomes more embedded in Lean processes, leaders must ensure the ethical use of technology. Prioritize transparency, fairness, and data privacy. Leaders need to balance technological advancement with corporate responsibility to earn the trust of both employees and customers.

Develop a Culture of Continuous Learning: In the evolving world of Lean 5.0+, continuous upskilling is essential. Leaders should create opportunities for their teams to learn and adapt to new technologies like AI and robotics. A learning-focused culture ensures that teams stay agile and can effectively integrate emerging innovations into Lean processes.

Prioritize Employee Empowerment: Lean 5.0+ emphasizes human-centered approaches. Leaders must empower employees by involving them in decisions on how AI and automation are implemented. By encouraging collaboration and creativity, leaders can maintain a workforce that feels valued and essential, even as automation expands.

Lead with Agility and Adaptability: The rapid pace of technological change requires leaders to adopt agile leadership styles. Embrace a mindset of flexibility and adaptability, being open to evolving business models, Lean processes, and AI integration strategies that optimize both human and technological contributions.

Integrate AI for Strategic Decision-Making: AI-driven insights are powerful tools for decision-making. Leaders should incorporate AI and data analytics to refine Lean processes, anticipate challenges, and seize opportunities. However, human intuition and leadership should guide the strategic use of AI to align with organizational values and goals.

Focus on Sustainability and Purpose: Lean 5.0+ aligns with Industry 5.0's emphasis on sustainability and societal impact. Leaders should integrate sustainable practices within their Lean strategies, focusing on reducing waste, conserving resources, and promoting ethical practices throughout the value chain.

Champion Human-AI Synergy: Leaders should focus on fostering collaboration between humans and AI. Recognize that AI enhances human capabilities rather than replaces them. By promoting AI to support employees, leaders can drive innovation and maintain the human touch in decision-making.

By adhering to these guidelines, leaders can effectively navigate the intersection of technology and humanity in Lean 5.0+, ensuring that AI enhances organizational efficiency while maintaining a strong human-centered ethos.

CONCLUSION

In conclusion, the transition from Industry 4.0 to Industry 5.0 marks a critical evolution in the integration of human and technological capabilities, reshaping the foundations of industries worldwide. As AI becomes more sophisticated and intertwined with human processes, Lean 5.0 emerges as a pivotal methodology, ensuring that Lean principles of waste reduction, process optimization, and customer value are adapted to this new era. Lean 5.0 goes beyond traditional digitalization, placing a human-centric focus on collaboration between AI, robotics, and the workforce, emphasizing the enhancement of human creativity, judgment, and innovation.

For organizations and leaders, the future will require embracing this synergy between technology and humanity, ensuring ethical AI integration while fostering a culture of continuous learning, adaptability, and sustainability. Lean 5.0 promotes a new leadership paradigm, one that balances technological advancements with empathy and humanistic values. Leaders who can navigate these dual forces – leveraging AI for strategic decisions while maintaining a focus on human empowerment – will be poised to lead their organizations to success in the age of AI-driven transformation. By aligning with the principles of Lean 5.0, organizations can thrive in this exponentially advancing industrial landscape.

REFERENCES

Dalal, A. (2024a). The AI revolution: Reshaping corporate culture. *LinkedIn*. https://www.linkedin.com/pulse/ai-revolution-reshaping-corporate-culture-dr-adil-dalal-ymysc/

Dalal, A. (2024b). Leadership in the age of AI: At the crossroads of humanity and technology. *Forbes*. https://www.forbes.com/councils/forbescoachescouncil/2024/09/09/leadership-in-the-age-of-ai-at-the-crossroads-of-humanity-and-technology/

Institute of Entrepreneurship Development. (2021). *The 4 industrial revolutions*. https://ied.eu/project-updates/the-4-industrial-revolutions/

Krüger, J. (2023). The 5th industrial revolution: The dawn of the cognitive age. *Psychology Today*. https://www.psychologytoday.com/us/blog/the-digital-self/202310/the-5th-industrial-revolution-the-dawn-of-the-cognitive-age

Nahavandi, S. (2019). Industry 5.0 – a human-centric solution. *Sustainability*, *11*(16), 4371. https://doi.org/10.3390/su11164371

Pine, B. J., II, Peppers, D., & Rogers, M. (1997). The four faces of mass customization. *Harvard Business Review*, *75*(1), 91–101. https://hbr.org/1997/01/the-four-faces-of-mass-customization

Schwab, K. (2016). *Industry 4.0: The fourth industrial revolution*. World Economic Forum.

Schwab, K. (2017). *The fourth industrial revolution*. Crown Publishing Group.

17 How the AI Revolution Will Transform the Corporate Ecosystems of the Future

THE AI REVOLUTION

AI is advancing at a rapid pace, reshaping corporate ecosystems and redefining operational and cultural norms. From transforming decision-making processes to reimagining workforce dynamics, AI is catalyzing a profound shift across industries.

REVOLUTIONIZING WORKFORCE DYNAMICS WITH INTELLIGENT AUTOMATION

AI's capacity to automate routine and repetitive tasks is redefining workforce roles and enhancing efficiency. Advanced AI systems allow employees to transition from mundane duties to strategic and innovative responsibilities. For example, Amazon's AI-powered robots optimize inventory management and order fulfillment, freeing human workers to focus on logistics strategy and customer service improvements (Kumar, 2020).

This transformation emphasizes the need for a workforce skilled in critical thinking, creative problem-solving, and interpersonal communication. A PwC report (2021) highlights that 77% of executives prioritize reskilling as a cornerstone of organizational success, fostering a culture of continuous learning and adaptability.

EMPOWERING DATA-DRIVEN DECISION MAKING

AI-powered analytics and decision-support systems are transforming organizations into hubs of evidence-based decision-making. These tools harness vast datasets to enable leaders and employees to make informed, objective, and strategic choices. Netflix's recommendation algorithm, for instance, uses data to deliver personalized content experiences, significantly boosting customer engagement and retention (Gomez-Uribe & Hunt, 2016).

Google exemplifies the cultural shift toward data-backed innovation by embedding analytics into its core values, promoting accountability, transparency, and strategic problem-solving (Sullivan, 2020).

DOI: 10.1201/9781003532972-17

Establishing Ethical AI Frameworks

The integration of AI into corporate operations presents ethical challenges, such as data privacy, algorithmic bias, and societal impacts. Organizations are prioritizing responsible AI governance to address these issues. Microsoft's AI Ethics Committee, for example, ensures that its AI solutions adhere to principles of transparency, fairness, and societal alignment (Holstein et al., 2019)

Such initiatives are reshaping corporate values, emphasizing trust and ethical responsibility as foundational elements of AI implementation (Binns, 2018).

Transforming Work Structures for Flexibility and Resilience

AI-driven tools are revolutionizing workplace efficiency by automating complex tasks, prompting a reassessment of traditional work structures. Salesforce's use of AI-driven scheduling tools optimizes meeting management, freeing employees to focus on high-value activities (Salesforce, 2022).

This shift fosters flexible work models, such as hybrid arrangements and adaptive hours, creating a culture that values employee well-being, productivity, and resilience.

Amplifying Creativity and Human-Centric Innovation

The automation of routine tasks underscores the unique value of human skills, such as creativity, emotional intelligence, and innovative problem-solving. 3M's "15% Time" policy exemplifies this shift, encouraging employees to dedicate a portion of their time to passion projects, resulting in groundbreaking innovations like Post-it Notes (Fry, 2019).

Organizations are investing in collaborative environments and tools to promote brainstorming and human-centric creativity, recognizing their critical role in maintaining a competitive edge.

Redefining Customer Interactions through Personalization

AI's ability to deliver personalized experiences at scale is revolutionizing customer engagement. Sephora's AI-powered chatbots provide tailored product recommendations, enhancing customer satisfaction and loyalty (Lee, 2020).

While AI addresses routine queries, human employees are trained to manage complex, emotionally nuanced interactions, fostering a culture that values both efficiency and high-touch service.

Promoting Diversity, Equity, and Inclusion through AI

AI offers a unique opportunity to mitigate bias in hiring and promotion processes by emphasizing objective criteria. For example, Unilever's AI-powered recruitment tools assess candidates based on skills and potential rather than subjective factors, advancing diversity and inclusivity (LinkedIn, 2020).

This commitment to fairness and equity fosters inclusive corporate cultures. By designing AI systems to identify and reduce biases, organizations are building equitable environments that support diverse talent.

Driving Corporate Agility and Innovation

AI enables organizations to adapt swiftly to market changes and foster innovation. Tesla's use of AI in product development and manufacturing allows for rapid prototyping and iterative improvements, ensuring responsiveness to evolving consumer demands (Sharma, 2021).

This agility supports a culture of continuous improvement, empowering companies to capitalize on emerging opportunities and navigate disruptions effectively.

Enhancing Strategic Leadership with AI Insights

AI is transforming leadership by providing executives with deeper insights into organizational performance and market dynamics. IBM's Watson Analytics equips leaders with predictive analytics, enabling data-driven decision-making and uncovering growth opportunities (IBM, 2021).

This shift fosters a leadership culture that prioritizes data literacy, adaptability, and ethical considerations, aligning with the demands of a rapidly evolving corporate landscape.

Revolutionizing Collaboration through AI-Enhanced Tools

AI-powered collaboration platforms are enhancing teamwork and cross-functional communication. Slack's integration of AI features, such as automated task prioritization and sentiment analysis, improves team productivity and cohesion (Slack, 2022).

These advancements are fostering a culture of seamless collaboration, where technology complements human interactions to achieve organizational goals efficiently.

Optimizing Supply Chains with Predictive Analytics

AI is revolutionizing supply chain management by predicting demand, optimizing inventory, and streamlining logistics. Walmart's use of AI-driven analytics ensures efficient stock management and minimizes waste, aligning operations with sustainability goals (McKinsey, 2021).

This integration of AI enhances operational efficiency, supporting a corporate culture that values sustainability and responsiveness.

CONCLUSION: BALANCING INNOVATION AND HUMANITY

The integration of AI into corporate ecosystems marks more than a technological evolution; it signifies a cultural transformation. By redefining workforce dynamics, decision-making frameworks, customer experiences, and corporate values, AI is fundamentally reshaping organizational paradigms.

Companies that navigate this transition successfully will achieve a harmonious balance between leveraging AI's capabilities and fostering human-centric attributes, such as creativity, emotional intelligence, and ethical leadership. Leading organizations like Amazon, Netflix, Tesla, and Unilever demonstrate how embracing AI's transformative potential can drive innovation, inclusivity, and long-term success in an increasingly competitive global landscape.

REFERENCES

Binns, R. (2018). Fairness in machine learning systems. *Microsoft Research.*

Fry, R. (2019). Innovation strategies at 3M. *Harvard Business Review.*

Gomez-Uribe, C. A., & Hunt, N. (2016). The Netflix recommender system: Algorithms, business value, and innovation. *ACM Transactions on Management Information Systems, 6*(4), 1–19.

Holstein, K., Wortman Vaughan, J., Daumé III, H., Dudík, M., & Wallach, H. (2019, May). Improving fairness in machine learning systems: What do industry practitioners need? In *Proceedings of the 2019 CHI conference on human factors in computing systems* (pp. 1–16). ACM. https://doi.org/10.1145/3290605.3300830

IBM. (2021). Predictive analytics with Watson. *IBM Analytics.*

Kumar, A. (2020). Automation at Amazon: The role of AI in supply chain management. *Journal of Supply Chain Innovation, 8*(2), 45–58.

Lee, S. B. (2020). Chatbots and communication: The growing role of artificial intelligence in addressing and shaping customer needs. *Business Communication Research and Practice, 3*(2), 103–111.

LinkedIn. (2020). AI in recruitment: Unilever's approach to diversity. *LinkedIn Talent Solutions.*

McKinsey. (2021). AI in supply chain management: Insights and case studies. *McKinsey Insights.*

PwC. (2021). Workforce of the future: The competing forces shaping 2030. *PwC Insights.*

Salesforce. (2022). AI-powered productivity tools. *Salesforce Blog.*

Sharma, R. (2021). AI-driven innovation at Tesla. *Journal of Technology Management, 12*(3), 34–47.

Slack. (2022). Enhancing collaboration with AI tools. *Slack Insights.*

Sullivan, J. (2020). Data-driven innovation at Google. *Forbes.*

18 The Detrimental Effects of AI on Learning, Decision-Making, and Cognitive Functions among Corporate Employees

Impacts on Learning and Skill Development

The automation of routine tasks by AI can lead to a decline in employees' opportunities to develop and refine essential skills. Delegating tasks such as data analysis, strategic planning, or customer service to AI systems reduces opportunities for employees to engage in problem-solving and critical thinking. Over time, this can result in skill degradation and cognitive disengagement.

Case Study: Aviation and Automation Reliance

In aviation, the implementation of intelligent decision support systems has contributed to skill degradation among pilots. Cummings (2004) highlighted that over-reliance on autopilot systems reduced pilots' situational awareness and manual flying skills. This reliance becomes problematic during unexpected scenarios requiring manual intervention, emphasizing the need for balanced human involvement alongside AI.

Additionally, the effects of skill atrophy extend beyond aviation. Industries heavily reliant on automated decision systems, such as manufacturing and logistics, report similar challenges. Employees tasked with overseeing automated systems often struggle with maintaining proficiency in manual interventions, further highlighting the widespread implications of AI dependency.

DECISION-MAKING CHALLENGES: AUTOMATION BIAS

Automation bias, the tendency to favor AI-generated recommendations without sufficient scrutiny, is a significant concern in AI-integrated workplaces. Employees often

DOI: 10.1201/9781003532972-18

trust AI outputs even when contradictory or incomplete information is available, potentially leading to errors.

Case Study: Healthcare Decision Support Systems
Clinical Decision Support Systems (CDSS) in healthcare exemplify automation bias. Lyell and Coiera (2017) found that clinicians frequently over-relied on CDSS, leading to errors of commission (following incorrect recommendations) and omission (failing to act without system prompts). This bias underscores the need for critical evaluation and human oversight in AI-assisted decision-making processes.

Case Study: Financial Services and Risk Analysis
In the financial sector, reliance on AI for risk analysis and investment decisions has demonstrated similar pitfalls. Analysts using AI tools often fail to recognize nuanced risks not accounted for in the algorithm, resulting in significant financial losses (PwC, 2021). This reinforces the importance of human judgment in complementing AI insights.

NEUROPLASTICITY AND COGNITIVE ENGAGEMENT

Neuroplasticity, the brain's ability to reorganize itself through learning and experience, is critical for cognitive health and adaptation. However, AI's role in automating mentally stimulating tasks can limit neural engagement, potentially hindering employees' cognitive development.

Case Study: Administrative Roles and AI
Klingberg (2019) explored the cognitive impact of AI automation on administrative assistants. Employees using AI tools for scheduling, task prioritization, and document preparation showed diminished problem-solving abilities over time. This decline highlights the risks associated with reduced cognitive engagement when AI takes over routine yet stimulating tasks.

Expanding this understanding, a broader analysis of AI's impact on neuroplasticity in creative roles, such as marketing and product development, suggests that employees who frequently depend on AI for ideation and content creation show reduced originality in their work. This underscores the importance of balancing automation with opportunities for creative input.

ETHICAL CONCERNS: DESKILLING AND EMPLOYEE DEPENDENCE

The widespread adoption of AI poses ethical challenges, particularly concerning deskilling and increased dependence on technology. Deskilling occurs when employees lose proficiency in core competencies due to prolonged reliance on AI. This dependence not only reduces individual growth but also weakens organizational resilience in the event of AI system failures.

Case Study: Radiology and Diagnostic AI
In radiology, over-reliance on AI diagnostic tools has led to a decline in manual diagnostic skills. Cummings (2018) observed that radiologists dependent on AI tools experienced a measurable reduction in diagnostic accuracy when required to analyze cases without AI support. This highlights the importance of maintaining human expertise alongside AI integration.

Furthermore, deskilling is a growing concern in industries such as retail and supply chain management, where AI systems dominate inventory control and logistics. Employees increasingly report challenges in adapting to manual systems during AI downtimes, illustrating the long-term risks of over-dependence.

PSYCHOLOGICAL IMPACTS: REDUCED CONFIDENCE AND MOTIVATION

AI's increasing role in workplaces can negatively affect employees' confidence and motivation. When employees perceive AI as superior in efficiency and accuracy, they may undervalue their contributions, leading to feelings of redundancy and reduced job satisfaction.

Case Study: Retail Sector and AI-Driven Inventory Systems
A report by Forbes Technology Council (2024, November 18) investigated the psychological effects of AI-driven inventory systems in retail. Employees expressed concerns about job obsolescence and reported decreased morale, with many feeling disengaged from their roles. This underscores the need for organizations to address the psychological impact of AI on their workforce.

Case Study: Technology Startups and Employee Burnout
Technology startups relying on AI-driven productivity tools often report higher employee burnout rates. Wiethof et al. (2022) found that employees working alongside high-efficiency AI systems often experienced heightened pressure to match AI outputs, contributing to stress and job dissatisfaction.

COLLABORATIVE PROBLEM-SOLVING AND TEAM DYNAMICS

AI's role in workplace collaboration tools can inadvertently weaken human teamwork and collective problem-solving skills. Automated project management platforms streamline communication but may discourage in-depth collaboration among team members, affecting innovation and creativity.

Case Study: Startups Using AI Collaboration Tools
Creativity and Innovation in the Age of AI
In the Forbes article "The V-Factor: Unlocking the Power of Visualization" (Dalal, 2025b), the V-Factor is presented as a critical tool for enhancing human creativity and innovation – capabilities that are increasingly essential in the age of AI. Visualization is framed not as passive imagination but as a deliberate cognitive process that stimulates neuroplasticity, rewires mental pathways, and fuels problem-solving. By envisioning goals with clarity and emotional intensity, individuals can activate the brain's creative centers, strengthen decision-making, and spark innovative thinking.

In a world where AI handles logic and data at scale, the V-Factor empowers humans to stay ahead by amplifying their innate strengths – imagination, intuition, and inventiveness. This practice turns abstract ideas into actionable insights and fosters mental resilience, helping individuals and organizations remain agile and adaptive. When embedded into daily routines, visualization becomes a catalyst for breakthroughs, especially in leadership, strategy, and design thinking. Mastering the V-Factor is not just beneficial – it's vital for anyone seeking to thrive in a rapidly evolving technological landscape where creativity remains humanity's competitive edge.

MITIGATING NEGATIVE IMPACTS: BEST PRACTICES AND RECOMMENDATIONS

Apart from the V-Factor, organizations can counteract AI's detrimental effects by implementing strategies that balance technology integration with employee development:

Continuous Learning Programs: Encourage employees to participate in training programs that develop skills complementary to AI capabilities (PwC, 2021).

Promoting Critical Thinking: Design workflows that require employees to validate AI outputs and engage in decision-making processes actively.

Encouraging Collaboration: Create opportunities for team-based problem-solving that leverage human creativity and AI's analytical power.

Monitoring Ethical AI Use: Establish guidelines to ensure AI complements human expertise rather than replacing it.

Providing Mental Health Support: Recognize the psychological impacts of AI integration and offer resources to help employees adapt and thrive.

CONCLUSION

While AI offers transformative potential for corporate environments, its over-reliance can have adverse effects on learning, decision-making, and cognitive functions among employees. This chapter underscores the risks of automation bias, deskilling, and cognitive stagnation through verified case studies and research. By adopting the power of visualization, and a balanced approach to AI integration, organizations can enhance productivity while preserving and fostering human cognitive and emotional strengths. Such strategies will ensure that AI serves as an empowering tool rather than a source of dependency and decline.

REFERENCES

Cummings, M. L. (2004). Automation bias in intelligent decision support systems. *Advances in Human Factors and Ergonomics*, *7*(4), 103–120.

Cummings, M. L. (2018). Human-automation collaboration in radiology. *Journal of Medical Imaging*, *5*(2), 215–230.

Dalal, A. F. (2025a). *The V-Factor: The power of visualization – Your ultimate key to neuro-excellence*. Pinnacle Process Solutions.

Dalal, A. (2025b). The V-factor: Unlocking the power of visualization. *Forbes*. https://www.forbes.com/councils/forbescoachescouncil/2025/04/15/the-v-factor-unlocking-the-power-of-visualization/

Forbes Technology Council. (2024, November 18). How spatial AI can help reshape retail experiences. *Forbes*. https://www.forbes.com/councils/forbestechcouncil/2024/11/18/how-spatial-ai-can-help-reshape-retail-experiences/

Klingberg, T. (2019). Neuroplasticity and cognitive engagement in the age of AI. *Cognitive Research*, *14*(2), 101–117.

Lyell, D., & Coiera, E. (2017). Automation bias and clinical decision support systems. *Journal of the American Medical Informatics Association*, *24*(6), 1156–1165.

PwC. (2021). *Workforce of the future: Reskilling strategies for AI integration*. https://www.pwc.com

Wiethof, C., et al. (2022). Toward a hybrid intelligence system in customer service: Collaborative learning of human and AI. *Journal of Service Management*, *33*(4), 67–80.

Index

FSC
www.fsc.org
MIX
Papier | Fördert gute Waldnutzung
FSC® C083411

Kolibri 360 GmbH
Ferdinand-Jühlke-Straße 7
99095 Erfurt, Deutschland
produktsicherheit@kolibri360.de